# TEACHING AGRICULTURE
## *Through*
# PROBLEM SOLVING

Third Edition

# TEACHING

**JOHN R. CRUNKILTON, Ph.D.**

*Professor and Program Area Leader
of Agricultural Education
Virginia Polytechnic Institute and
State University
Blacksburg*

**ALFRED H. KREBS, Ph.D.**

*Vice President of Administration, Emeritus
Virginia Polytechnic Institute and
State University
Blacksburg*

# AGRICULTURE
## *Through*
# PROBLEM SOLVING

Order from

**The INTERSTATE PRINTERS & PUBLISHERS, *Inc.***

Danville, Illinois 61832-0594

**TEACHING AGRICULTURE THROUGH PROBLEM SOLVING,** Third Edition. Copyright © 1982 by The Interstate Printers & Publishers, Inc. All rights reserved. Prior editions published under the title *For More Effective Teaching* in 1954 and 1967. Printed in the United States of America.

Library of Congress Catalog Card No. 81 84478

ISBN 0-8134-2199-3

To our wives,

**Sherry S. Crunkilton**

and

**Jean E. Krebs,**

and to

**Richie Crunkilton,**

in hopes that his teachers use
problem solving in helping him to solve problems

# Foreword

*T*HIS BOOK WAS WRITTEN TO ASSIST TEACHERS *in their approach to the task of guiding the learning activities of their students. We make this effort to explain and demonstrate how problem solving can lead to an effective teaching / learning situation because this approach to teaching will help teachers address the problems facing their students. Teaching can be one of the most satisfying and rewarding of careers if the teacher experiences success in the classroom and laboratory. Without this success, the teacher will be frustrated and unhappy, and this may result in a negative influence on other areas of the educational program as well as on the teacher's personal life.*

*The successful use of problem solving will make school more interesting, exciting, and enjoyable for both the student and the teacher.*

JRC<br>AHK

# ▶ Acknowledgments ◀

*T*HE AUTHORS WISH TO RECOGNIZE THOSE INDIVIDUALS *who provided suggestions and comments for the original edition of the book,* **For More Effective Teaching.** *They are Dr. G. Fuller, Dr. P. E. Hemp, Dr. L. J. Phipps, Dr. J. R. Warmbrod, and Mr. Arthur B. Ward. From that original book, ideas came forth for this revision.*

*Special appreciation also goes to the hundreds of teachers in Illinois and Virginia who have given us ideas for clarifying and refining our thoughts on the teaching / learning process.*

JRC
AHK

# Contents

|  | PAGE |
|---|---|
| *Foreword* | vii |
| *Acknowledgments* | ix |

CHAPTER

1 **Understanding Planning** ............................. 1

    *The parts of the plan for teaching using
a problem-solving approach, 4
Enterprise or activity, 6
Unit versus problem area, 6
Teaching procedures, 16
Teaching resources, 28*

2 **Theory and Practice** ................................. 33

    *Some principles of learning, 35
Some concepts about learning, 41
Some related studies, 55*

3 **Motivating Students** ................................. 69

    *Aspects of motivation, 72
Using questions in motivating students, 80*

4 **Selecting Teaching Techniques** ....................... 85

    *Teaching strategy, 88
Teaching techniques, 89
Instructional aids, 89
Teaching techniques and problem solving, 90*

| CHAPTER | PAGE |
|---|---|

    *Functions of teaching techniques and instructional aids, 90*
    *Deciding which teaching techniques to use, 94*

**5 Reviewing Some Illustrative Plans** ...................... **105**

    *Plan number one (a problem area plan), 107*
    *Plan number two (a problem area plan), 109*
    *Plan number three (a unit plan), 112*
    *Plan number four (a problem area plan), 116*
    *Plans in agricultural mechanics, 120*
    *Plan number five (a problem area plan), 122*
    *Plan number six (a unit plan), 125*
    *Plan number seven (a problem area plan), 127*
    *Teaching "new developments" units, 130*
    *Adjustments for agribusiness occupations instructors, 131*

**6 Taking the Plan into the Classroom** ..................... **135**

    *What the teacher does in the classroom, 137*
    *Conducting laboratory activities, 144*

**7 Avoiding Trouble Spots** ................................. **145**

**8 A Minimum Plan** ....................................... **157**

    *A minimum plan, 159*
    *The minimum plan illustrated, 161*

**9 Variations in Using Problem Solving** .................... **167**

    *Variation number one, 169*
    *Variation number two, 171*
    *Variation number three, 171*
    *Variation number four, 173*
    *Variation number five, 173*

**10 Using Problem Solving to Control Discipline** ............. **175**

    *Meaning of discipline, 177*
    *To obtain good discipline, 178*
    *To correct poor discipline, 180*

| CHAPTER | PAGE |

    *Measures to avoid, 181*
    *Deciding on the controls, 182*
    *Deciding when physical force is justified, 184*
    *Summary, 184*

**11 Practices for conducting an Effective Agricultural Laboratory Program** ........................ 187

**12 Using Problem Solving in Teaching Adults** ............... 203

    *Outline of the teaching plan for adult instruction, 207*
    *Taking the plan into the classroom, 213*
    *Illustration number one–crop production, 213*
    *Illustration number two–indoor plants, 215*
    *Illustration number three–beef production, 217*
    *Illustration number four–farm management, 218*
    *A special word on young farmer programs, 226*

**13 Evaluation of Teaching / Learning Activities** ............... 229

    *Analysis of teaching techniques, 231*
    *Evaluation of teaching, 241*
    *Analysis of the teacher, 245*
    *Laboratory instruction, 246*

**14 The Teacher Is the Key** ................................. 249

| APPENDIX | PAGE |

**A Building Courses of Study for Agricultural Education** ............................... 255

    *Steps in building a course of study, 257*
    *Principles or guidelines for course building, 258*

**B Unit and Problem Area List** ........................... 263

    *Units and problem areas, 265*
    *Agricultural business course, 289*
    *Adjustments to local situations, 293*

APPENDIX PAGE

C  Using a Problem-Solving Approach
   in Other Teaching Situations .......................... 295

   *Plan number one, 297*
   *Plan number two, 299*
   *Plans for other areas, 300*

D  Using and Developing "Principles" in
   Teaching Agriculture ................................. 307

E  Process Outline for Giving a Demonstration ............. 315

   *Bibliography* ........................................ 319
   *Index* .............................................. 325

# ▶ List of Figures ◀

| FIGURE | | PAGE |
|---|---|---|
| 2-1 | Relationship of Practice to Learning | 37 |
| 2-2 | Satisfying and Annoying Aspects of the Teaching / Learning Process | 39 |
| 4-1 | Selected Action Verbs Appropriate to the Domains of the Taxonomy: Cognitive | 97 |
| 4-2 | Selected Action Verbs Appropriate to the Domains of the Taxonomy: Affective | 98 |
| 4-3 | Selected Action Verbs Appropriate to the Domains of the Taxonomy: Psychomotor | 99 |
| 11-1 | Laboratory Clean-up Schedule | 198 |
| 13-1 | Analysis of Teaching Techniques | 232 |
| 13-2 | Behavioral Emphasis of Teaching Techniques | 239 |
| 13-3 | Were the Instructional Objectives Achieved? | 240 |
| 13-4 | Evaluation of Teaching | 242 |
| 13-5 | Student Participation | 244 |
| 13-6 | Student Participation Influenced by Teacher Stance | 245 |

◆ **CHAPTER** ◆

# 1

# ◆ Understanding Planning ◆

THE SUCCESS OR FAILURE OF TEACHING often can be traced directly to the effectiveness of planning. The value of well-constructed plans cannot be overestimated. However, to make a plan just for the sake of having a plan is not enough. The plan must be usable. It must work when the teacher uses it in the classroom. It should reflect both what the teacher needs to do to prepare for teaching and what will happen when the plan is put into operation.

The suggestions presented here for making teaching plans do reflect what the teacher needs to do to prepare for teaching and what happens when the plan is used in the classroom. The complete plan is a combination of those two elements. The best things which can be said of the procedure, however, are that it works and that it is easy to understand. Of course, the teacher must be willing to take the trouble to make good plans, to stay with the procedure long enough to develop the ability to use it, and to acquaint the students with their responsibilities under this particular method of teaching.

A few persons have the mistaken impression that the use of this particular approach to teaching is limiting in terms of the various teaching techniques which may be used. Nothing could be further from the truth! This approach to teaching provides a framework within which any teaching technique may be used. It provides the teacher with a plan for organizing teaching which includes the maximum freedom for using the teacher's own creative talents to teach in a manner that will be most successful and satisfying both to the teacher and to the student. Even the lecturer would perform better if the lecture were to be presented from an outline (described in this text) consisting of the parts of the teaching plan utilizing problem solving. Then the emphasis and approach to content would be from the perspective of the student.

The following is the outline of a complete plan. It is very simple to make, but proper use of the plan will be a major stride toward the effective use of a problem-solving approach in teaching. Some teachers prefer to organize their work in units rather than in problem areas. For maximum clarity, separate outlines are provided for organizing from a problem area approach and from a unit approach even though there is little difference between the two in the overall procedure.

## THE PARTS OF THE PLAN FOR TEACHING USING A PROBLEM-SOLVING APPROACH

The plan for a problem-solving approach to teaching is presented in three distinct parts. Part I consists of the analysis the teacher should perform prior to planning for the classroom teaching experience. Part II consists of the plan for that which is to take place in the classroom. Part III is a listing of the teaching resources to be used.

### Problem Area Plan

#### Part I. Pre-planning Analysis
A. Selection of the content—enterprise or activity and problem area (content segment)
B. Analysis of the teaching situation
C. Statement of the teaching / learning objectives

#### Part II. The Teaching Plan
A. Enterprise or activity
B. Problem area (content segment)
C. Interest approach
D. Anticipated group objectives
E. Anticipated problems and concerns
F. Steps in solving the problems and concerns
G. Evaluation and application

#### Part III. Teaching Resources
A. References
B. Teaching aids
C. Teaching techniques

## Unit Plan

### Part I. Pre-planning Analysis
A. Selection of the content—enterprise or activity and unit
B. Analysis of the teaching situation
C. Statement of the teaching / learning objectives

### Part II. The Teaching Plan
A. Enterprise or activity
B. Unit:
Problem areas (content segments) in unit:
1.
2.
3.
C. Interest approach for unit
D. Anticipated group objectives for unit
E. Study of problem areas
   1. Problem area #1
      a. Discussion of unit objectives which apply
      b. Anticipated problems and concerns
      c. Steps in solving the problems and concerns
      d. Listing approved practices
   2. Problem area #2
      a. Discussion of unit objectives which apply
      b. Anticipated problems and concerns
      c. Steps in solving the problems and concerns
      d. Listing approved practices
   3. Problem area #3
      a. Discussion of unit objectives which apply
      b. Anticipated problems and concerns
      c. Steps in solving the problems and concerns
      d. Listing approved practices
F. Evaluation and application for the unit

### PART III. Teaching Resources
A. References
B. Teaching Aids } for each problem area
C. Teaching Techniques

## Understanding the Plan

It has often been said that one teacher cannot make a teaching plan for another teacher that will truly satisfy the other teacher. This is true if the second teacher does not have a complete understanding of the plan and of the approach to teaching for which the plan was designed. The following description of the parts of the plans outlined previously is intended to bring about this understanding.

## ENTERPRISE OR ACTIVITY

This indicates the broad heading under which the unit or problem areas to be studied may be found. Using the enterprise or activity heading keeps the teacher and student oriented to the general area of study. In most cases, the enterprise or activity is the broadest topical breakdown in the course of study.

## UNIT VERSUS PROBLEM AREA

The major difference between the problem area plan and the unit plan lies in the fact that the unit is made up of several problem areas. Thus, in the unit plan, the situation, teacher objectives, interest approach, anticipated group objectives, and evaluation must be developed for the unit as a whole. The unit plan also involves the additional steps of selecting the problem area to be considered and relating unit objectives to the appropriate problem area. The rest of the plan and the procedures for using the plan are the same for both approaches.

### Unit

The unit, if used, is the second major breakdown in the course of study. Units are used for grouping problem areas which have elements in common for convenience in teaching and studying. A unit should be limited to an area of content which can be taught without the need for interrupting to insert other areas of instruction in order to maintain interest. For the most part, a unit should be small enough so that the instruction about it can be accomplished in less than three weeks. Many students begin to lose interest when the instruction on one unit lasts for too long a period of time.

## Problem Area (Content Segment)

The problem area indicates the specific area for study within a particular enterprise, activity, or unit. It is called a problem area because it includes a great variety of problems, all of which relate to the specific area of subject matter under consideration.

The proper determination of the problem area is a critical point in the use of the problem-solving approach which is presented here. Each problem area should be a meaningful whole, with a minimum of overlapping with other problem areas. If the problem area is not properly determined, the problem-solving approach will not be effective. Problem areas which are too narrow limit the amount of analyzing which can be done by the students; problem areas which are too broad result in confusion when the pupils attempt analysis.

Problem areas can be smaller in scope when content is organ-

TEACHING MUST BE RELATED TO LOCAL CONDITIONS.

ized into units than when it is organized into problem areas only. The reason for being able to use smaller problem areas with a unit organization is that the unit provides the breadth needed for challenging the students' abilities and, at the same time, serves to indicate the limits within which the analysis of content is to take place.

Most units and problem areas should be stated in terms of the kinds of activities people use in describing their work and in terms of the way in which their jobs are divided for performing the work. This is best known as a functional approach. To the extent possible, the units and problem areas should also be kept seasonal in scope to permit seasonal planning as well as seasonal doing.

The relationship of the instructional program to the farming programs, to placement-employment in nonfarm agricultural business programs, and to other forms of agricultural experience program activities is very important. Ideally, most of the instruction should lead to application in the agricultural experience programs of the students concerned.

### Examples of Units and Problem Areas[1]

The following are examples of units and problem areas. Not all teachers will agree with those listed here. For one thing, the importance of a particular enterprise in a particular community will affect the determination of the units and problem areas. For another, ideas differ as to what should be included in any particular problem area. If the general principles are adhered to, however, there will be no difficulty in developing suitable units and problem areas for teaching.

**Enterprise: Swine**

Unit A: Selecting and buying swine.
Problem areas:
1. Selecting swine by physical appearance.
2. Using records in selecting swine.
3. Buying swine.

Unit B: Feeding and caring for swine from breeding to weaning.
Problem areas:
1. Feeding and caring for the sow and gilt from breeding to farrowing.

---

[1] A more complete list of units and problem areas can be found in Appendix B.

## UNDERSTANDING PLANNING

2. Feeding and caring for the sow and litter at farrowing time.
3. Feeding and caring for the sow and litter from farrowing to weaning.

Unit C: Feeding and caring for swine after weaning.
   Problem areas:
   1. Feeding growing hogs for market and for herd replacements.
   2. Feeding and caring for the herd boar.

Unit D: Housing and equipping the swine enterprise.
   Problem areas:
   1. Providing housing for swine.
   2. Providing equipment for swine.

Unit E: Controlling diseases and parasites.
   Problem areas:
   1. Preventing and controlling diseases of swine.
   2. Controlling parasites of swine.

Unit F: Marketing and improving the swine enterprise.
   Problem areas:
   1. Marketing swine.
   2. Improving the swine breeding program.

Unit G: New developments.
   Problem areas:
   1. Reviewing swine production practices.
   2. Reviewing swine marketing practices.

**Enterprise: Beef**

Unit A: Selecting and buying beef.
   Problem areas:
   1. Selecting beef animals by physical appearance.
   2. Using records in selecting beef animals.
   3. Buying beef animals.

Unit B: Feeding and caring for the beef breeding herd.
   Problem areas:
   1. Feeding and caring for the breeding herd during gestation.
   2. Feeding and caring for the cow and calf.

Unit C: Feeding and caring for growing and finishing beef animals.
   Problem areas:
   1. Raising the beef heifer and young bull.
   2. Feeding and caring for finishing beef animals.

Unit D: Housing and equipping the beef enterprise.
  Problem areas:
  1. Providing housing for the beef enterprise.
  2. Providing equipment for the beef enterprise.
Unit E: Controlling diseases and parasites of beef.
  Problem areas:
  1. Preventing and controlling diseases of beef.
  2. Controlling parasites of beef.
Unit F: Marketing and improving the beef enterprise.
  Problem areas:
  1. Marketing the beef enterprise.
  2. Improving the beef breeding program.
Unit G: New developments.
  Problem areas:
  1. Reviewing beef production practices.
  2. Reviewing beef marketing practices.

**Enterprise: Corn**

Unit A: Selecting and buying seed corn.
  Problem areas:
  1. Determining the variety to grow.
  2. Buying seed corn.
Unit B: Planting operations for corn.
  Problem areas:
  1. Applying fertilizer at planting time.
  2. Applying chemicals for weeds and insects at planting time.
  3. Adjusting the planter and planting corn.
Unit C: Caring for the growing corn crop.
  Problem areas:
  1. Controlling weeds in the growing corn crop.
  2. Controlling insects in corn.
  3. Preventing and controlling diseases in corn.
Unit D: Harvesting, storing, and marketing corn.
  Problem areas:
  1. Harvesting the corn crop.
  2. Drying the corn crop.
  3. Storing the corn crop.
  4. Marketing the corn crop.
Unit E: New developments.
  Problem areas:
  1. Reviewing corn production practices.
  2. Reviewing corn marketing practices.

## Enterprise: Forage crops

Unit A: Establishing pastures.
 Problem areas:
 1. Selecting seed mixtures for pastures.
 2. Establishing permanent pastures.
 3. Establishing rotation pastures.

Unit B: Maintaining pastures.
 Problem areas:
 1. Maintaining permanent pastures.
 2. Renovating permanent pastures.

Unit C: Hay.
 Problem areas:
 1. Selecting seed mixtures for hay crops.
 2. Planting grasses and legumes for hay.
 3. Controlling insects and diseases of hay crops.

Unit D: Harvesting, storing, and marketing hay.
 Problem areas:
 1. Harvesting and storing hay.
 2. Marketing hay.
 3. Producing and marketing seed for hay crops.

Unit E: Producing corn silage.
 Problem areas:
 1. Growing corn for silage.
 2. Harvesting and storing corn for silage.

Unit F: Producing grass silage.
 Problem areas:
 1. Growing grasses and legumes for silage.
 2. Harvesting and storing grasses and legumes for silage.

Unit G: New developments in forage crops.
 Problem areas:
 1. Reviewing pasture production practices.
 2. Reviewing hay production, harvesting, and marketing practices.
 3. Reviewing corn silage production practices.
 4. Reviewing grass silage production practices.

## Enterprise: Soils and fertilizers

Problem areas:
1. Taking soil samples.
2. Making and using soil tests.
3. Determining fertilizer needs through deficiency symptoms and tissue tests.
4. Selecting and buying fertilizers.

5. Preparing the soil for planting.
6. Maintaining the organic matter content of the soil.
7. Handling and storing manure.
8. Determining soil capability classification.
9. Making a soil conservation plan.
10. Planning the long-time fertilizer program for the farm.
11. Reviewing new developments in soil fertility and management practices.

**Enterprise: Agricultural Business**

Unit: Selling agricultural products and services
Problem areas:
1. Selling techniques for agricultural products.
2. Selling techniques for agricultural services.
3. Operating business machines and equipment.
4. Writing up the sale.

## Other Factors to Be Considered

There are many decisions which have to be made in determining what the units and problem areas should be and where the individual units and problem areas should be placed in the overall plan or course of study. Some teachers may prefer to use the kind of organization of content indicated by the units and problem areas given for swine, beef, and corn. Other teachers may prefer the unit and problem area organization used for forage crops, where several crops are under one enterprise. Still other teachers may wish to use only a problem area organization much like that used for soils and fertilizers, or the problem area list which would remain if all "unit" headings were simply deleted from the others.

Determining the enterprise or activity under which a unit or problem area should be placed may also be a problem. Drainage and irrigation could be listed under agricultural mechanics or under soils and fertilizers. Soil conservation could be treated under soils, agricultural mechanics, a separate soil conservation enterprise heading, or the individual crops. Soil conservation could also be subdivided to fall in part under each of several enterprise headings. The placement of the unit and problem area *is* important. The proper determination of the unit and problem area is of even greater importance with the proper determination of the problem area being the most critical.

The illustrations just given should provide an adequate picture of the kinds of units and problem areas into which the various enterprises can be divided. In general, they follow the same line of thought one would follow in thinking about performing the tasks involved in each of the units and problem areas. The units and problem areas on a review of production and marketing practices serve to refresh the memories of the advanced students regarding past instruction while providing an opportunity to bring them up to date with the latest technical developments.

Nearly all of the technical agriculture which is to be taught can be approached from the standpoint of seasonal activity. An illustration of this is the approach to be used in teaching about tilth of the soil. This has been designated as a problem area by some teachers. It is much more effective and meaningful to consider tilth in connection with a problem area on preparing the soil for planting than it is to consider it as a problem area by itself. Too often, the teacher who refuses to organize content by problem areas, as just defined, wishes to organize content in the same way college courses are organized. This would make it possible for the teacher to use college notes and texts rather than to become involved in the difficult task of organizing content so that it will have the greatest meaning to the students.

In placing the units and problem areas into the course of study, full advantage should be taken of the seasonal activity method of determining units and problem areas. Those units and problem areas which have seasonal implications should be placed into the course of study or course calendar first; those of a non-seasonal nature can be placed into the course of study at times when there is less pressure to provide instruction for some seasonal activity.

### Situation

The situation includes certain information about the class being taught, about local agricultural conditions, and about state and national agricultural conditions when these are important to the total picture. The teacher needs to consider carefully the total situation in the light of each unit and problem area. The following items are typical of the kind of information which needs to be obtained when analyzing the teaching situation in preparation for developing the teaching plan:

1. The extent to which the students in the class are concerned

with the unit or problem area in their supervised agricultural experience programs.
2. The amount of instruction which the class has already received which relates either directly or indirectly to the unit or problem area.
3. The extent to which the unit or problem area is of community-wide concern.
4. The practices being followed in the community as related to the unit or problem area.
5. The state and national situation as related to the unit or problem area, whenever pertinent.
6. The timeliness of the instruction as it relates to work being done or work which is about to be done.
7. Production data.
8. Titles of nonfarm agricultural occupations in which the content of the unit or problem area would be useful.
9. Resources available for teaching.

There are undoubtedly other points which individual teachers will wish to include under the situation. The important thing to remember is that the teacher is not ready to continue with planning until the total situation has been adequately analyzed in relation to the members of the class. Of particular importance is the fact that the analysis of the situation helps to point up the teaching / learning objectives.

**Teaching / Learning Objectives**

Teaching / learning objectives are at the heart of any good teaching plan. They are the basis for the evaluation of both teaching and learning.

*Teaching / learning objectives*, as the term is used here, means only a simple list of the abilities the students should develop as a result of their studies. The teaching / learning objectives can be stated in terms of measurable objectives in order to be compatible with the recent changes in strategies for the evaluation of teaching and learning. If the objectives are sufficiently specific and narrow in focus, they can be used to develop individualized instruction or self-study materials.

Teaching / learning objectives are the point at which the interests of the teacher and those of the students come together. The students need to develop the abilities identified as teaching /

UNDERSTANDING PLANNING 15

learning objectives in order to accomplish their objectives which are often stated as production goals. The teacher will evaluate the effectiveness of the instruction given in terms of the extent to which the students develop the abilities stated as "teaching / learning objectives."

For an illustration, consider the teaching / learning objectives for the problem area, "Making and using soil tests."

The objectives could be stated broadly as follows: To develop the ability of the students to make and use soil tests in connection with their supervised agricultural experience programs, whether on a farm or in a nonfarm agricultural business.

BOTH TEACHER AND STUDENTS NEED CLEARLY DEFINED OBJECTIVES.

This, however, is not enough. It doesn't give the teacher much of a guide for planning or for teaching. More specific objectives are needed. For example, this broad objective would include

many specific contributory objectives such as developing the following abilities:
1. To determine what soil tests should be made and by whom.
2. To make the necessary soil tests.
3. To make soil maps showing the soil test results.
4. To interpret soil test results.
5. To relate soil test results to the fertilizer needs of the various crops.
6. To use and care for the equipment needed for testing soil.
7. To make proper use of the soil testing bureaus in the area.
8. To determine the relationship of soil test results to the various types of soil.
9. To understand the relationship between soil testing and efficiency in farming.
10. To understand the limitations of soil tests.
11. To determine how often to test each field.

Each teacher will wish to add to and subtract from this list of objectives. It is illustrative only. However, the better the job done at this point in planning, the better will be the rest of the plan and the teaching which follows. If a particular objective is to be accomplished, that objective must be recognized and the teaching organized so that it can be accomplished. In the final analysis, the teacher must evaluate teaching in terms of the teaching / learning objectives established.

## TEACHING PROCEDURES

### Interest Approach

The interest approach indicates the manner in which the teacher is going to introduce the unit or problem area to the class. This may be done through discussion, through the use of a film, by a field trip, by the use of specimens, or in some other way. The variety of approaches is limited only by the teacher's imagination. Whatever approach is used should be outlined carefully in the plan. It should result in the students' recognition of the problems involved in the unit or problem area as those they face in their supervised agricultural experience programs on a farm or in nonfarm agricultural businesses. The approach should stimulate a certain amount of enthusiasm in the class for studying the unit or problem area.

# UNDERSTANDING PLANNING

EXCHANGE OF EXPERIENCES IS ONE OF THE MOST VALUABLE PARTS OF CLASSROOM TEACHING.

The most common interest approach used is that of bringing the students into a general discussion of the unit or problem area. In this case, the plan for the interest approach would be a list of the leading questions to be asked. The leading questions should draw on the students' experiences and develop a picture of the situation similar to that which was outlined under the analysis of the situation. The emphasis should be on the supervised agricultural experience program situations.

When field trips, movies, and the like are used in developing the interest approach, the discussion which follows should center around what was observed as it relates to the supervised agricultural experience programs.

The interest approach serves several functions, some of which have been mentioned previously. The following list will serve to highlight those which are most important.

1. It arouses the interest of the students in the unit or problem area.

2. It helps the students recognize the problems as their own.
3. It helps the teacher obtain a fairly accurate picture of how much the students already know.
4. It develops a common background of information within the class regarding the unit or problem area being discussed.
5. It sets the stage for the establishment of student objectives.

A more complete discussion of motivation is presented in Chapter 3.

## Anticipated Group Objectives

The teacher should also list the reasons why the students should study the problems in a particular unit or problem area. This list will give the teacher a guide to follow in drawing from the students their reasons for studying. This section could well be included as a part of the interest approach since a portion of motivation depends on the students' understanding of why they should study the problems in a particular unit or problem area.

Many of the student objectives will be stated in the form of production goals. Other objectives will be stated in terms of other outcomes of doing well that which is indicated by the unit or problem area title. It is very important that these objectives be drawn out in such a way that they are recognized by both the students and the teacher.

The major question or questions the teacher plans to use to lead the students in developing objectives should also be a part of the plan.

## Anticipated Problems and Concerns of the Students

The teaching plan should include a list of the problems which the teacher anticipates will be raised by the students. This list serves several purposes:
1. It serves as a guide to the teacher in leading the discussion.
2. It helps the teacher make certain that all important problems have been raised.
3. It gives the teacher additional guidance in providing suitable references and other teaching materials.

# UNDERSTANDING PLANNING

4. It points up the field trips, demonstrations, and other special events for which the teacher needs to prepare.

As with the interest approach, the major questions the teacher plans to use in guiding the students as they identify their problems should be listed in the plan.

As a check to determine whether the list of problems and concerns is complete, the teacher should compare the problems list with the teaching / learning objectives list. If these two lists are supportive of each other with respect to all items, then the teacher can feel confident that the plan includes all needed content. If, however, there are no teaching / learning objectives for some of the problems, or if there are teaching / learning objectives for which there are no problems, then more planning is needed.

## Steps in Solving the Problems

The following steps to be taken to solve the problems will be the same for each plan and can be omitted when the teacher no longer needs them as a guide. A brief discussion of each step is included here in order to avoid any misunderstanding as to the procedure.

1. *Have the students select a problem from the list.*— Ordinarily, it doesn't make a great deal of difference which problem the students select first. However, if it does make a difference, then it is the teacher's responsibility to lead the students through the process of thinking which will result in the selection of the problem which should be considered first. If the decision is based only on the need for taking up certain of the problems in order to be ready for a field trip scheduled for a certain day, then the class should be advised of that fact and the problems taken up in the order required.

   If the students select a problem which is too limited, or which is a part of another problem, the teacher should point this out and suggest that certain problems be grouped and considered together.

2. *Lead the students in a discussion of the problem to find out what they do and do not know.*—At first, this may seem to be a simple matter. A more intensive examination, however, shows that this discussion is an analysis of the prob-

lem selected—an analysis which must be made carefully and thoroughly. A suggested approach is as follows:
   a. Ask the class, or the student who raised the problem, to explain the meaning of the problem.
   b. Ask the class to explain the relationship between this problem and the student objectives to point up the fact that the solution of the problem will contribute to the attainment of the objectives. Any misunderstandings regarding the objectives can be cleared up at this time.
   c. Lead the students in a discussion of how they are meeting this problem. Points of disagreement and points about which the students are in error are then listed on the board as a guide for the supervised study period. In order to keep the entire picture before the group, that which is known may also be listed on the board.

3. *Conduct supervised study on those things the students do not know or do not agree upon.*—The teacher should provide a great deal of intensive individual assistance during the supervised study period and should check frequently with the students to make sure they understand and are on the right track. Other teaching techniques may be used to assist the students in acquiring the technical information needed. (See Chapter 4 for additional suggestions on techniques to use.)

4. *Lead the students in the final discussion and drawing of conclusions.*—Here, the students present their individual findings for the points which were not known, and a group conclusion is drawn for each of the points for which information was sought in the reference materials. Then the entire problem is examined in the light of the new information, and the group decides whether or not the problem is solved sufficiently well. If so, the teacher then initiates study of the next problem. If not, additional work must be done on the problem until a suitable solution is found.

It is especially important at this point to make certain that the students understand *why* the solution to a problem is correct. It is the understanding of the *why*—the basic principles involved in the solution to a problem—which can be transferred to other problems in other situations.

5. *Have the students select another problem from the list and repeat the procedure.*—The process described in steps 1, 2, 3, and 4 is repeated until all of the problems for a problem area have been considered.

The same basic steps are followed for a unit plan. A problem area must be selected before student problems are identified and defined. Then, after solving all of the problems for the first problem area selected, it is necessary to return to the list of problem areas for the unit to determine the next problem area to study. After the next problem area is selected, it is analyzed into problems, and the problems are solved following the procedure already described.

It should be noted that reference is made to the unit objectives which apply before each problem area is analyzed into problems.

## Evaluation and Application

This part of the plan could just as well be called evaluation only, but the title used does serve to remind the teacher that the results of teaching are, in part, measured in terms of the extent to which the students put that which is learned into practice.

*Purposes.*—The purposes for which evaluation of teaching / learning is carried out are varied and numerous. Some examples of purposes are:
1. To determine the extent to which the teaching / learning objectives have been achieved.
2. To obtain data for determining grades earned by the students.
3. To identify areas of content needing additional emphasis.
4. To identify problems of individual students in understanding content.
5. To determine how much students know about a particular topic before beginning instruction on it.
6. To stimulate student interest in an area of content.
7. To identify individual student problems when application is made of the learnings in the world of work.
8. To assess the impact of instruction on individual student personal development, including maturing into adulthood.

*Techniques for evaluation.*—Each evaluative technique or procedure may be useful in accomplishing several of the purposes of evaluation. The most commonly used techniques are as follows:

1. *A list of approved practices.*—At the conclusion of the study of each problem area, it is good practice to draw from the students a list of all of the approved practices which have been developed through the study of the various problems. The students should understand that evaluation of their work will be based partly on the extent to which these approved practices are put into operation in connection with their supervised agricultural experience programs. Developing this list will serve as good review and will lay the foundation for the individual planning to follow.

   The list of approved practices should be prepared in advance as part of the teaching plan.

2. *Plans for supervised agricultural experience programs.*—If

CLASSROOM INSTRUCTIONS SHOULD BE BASED ON NEEDS GROWING OUT OF BROAD AGRICULTURAL EXPERIENCE PROGRAMS.

the teacher wants the classroom learning to be put into practice, the students must be given adequate opportunity to plan for this. All the teacher has to do in making teaching plans is to list this item so that it will not be forgotten. The teacher should not forget that all approved practices do not apply equally well to all situations. Each student must select those practices which can be put into effect in the student's own situation.

3. *Instruction and supervision on the job.*—This is another item for which no preparation is required in the teaching plan. It is listed only to remind the teacher that instruction is not completed, but only begun, in the classroom. Individual instruction on the job should be the teacher's most effective approach. Items 1 and 2 establish the foundation for this individual instruction on the job. Separate plans for on-the-job instruction should be developed.

4. *Test.*—Many tests are "made up" as the teacher writes the questions on the blackboard. Including the test as a part of the teaching plan insures at least a minimum amount of attention to the development of a good examination. As stated earlier, the test should measure the extent to which the teaching / learning objectives have been accomplished. For example, a test on "Making and Using Soil Tests" might include the following:
    a. A map of several fields on a farm—the students to indicate how they would take the soil samples.
    b. A map of a field showing the test results—students to explain how this information could be used in planning to fertilize the field for corn.

    It must be kept in mind that the classroom test is a poor technique for evaluation when compared with the other techniques available; namely, making plans for agricultural experience programs and on-the-job supervision to determine the extent to which the plans are put into effect.

## Criteria for Guiding Test Construction

There are certain criteria of value in guiding the teacher in test construction. The following are important considerations:

1. Is the subject matter adequately sampled? It is fairly easy to "trip up" even good students if only one or two points are selected for the test from a large body of content. By chance, the poor student may do better than the good student.
2. Does the test measure student accomplishment? The teaching / learning objectives may be the development of motor skills, the ability to organize, an understanding of terms, the ability to identify principles, or the ability to apply learning in problem situations.
3. Does the test discriminate among the students? This requires a balance among difficult questions, moderately difficult questions, and easy questions.
4. Does the test lend itself to consistency in scoring so that the grade derived from it would be about the same each time the same paper was scored?
5. Is the test economical of time-use in construction and scoring?
6. Does the test make guessing difficult?
7. Does the test parallel the work the students have been doing in class? The test should cover the work of the class if the purpose is to find out how well the students have done on a particular unit.
8. Is the test of the correct length? The test should be short enough to permit the students to answer each question to the best of their ability in the time allotted unless speed of response is one of the teacher's objectives.

## Specific Suggestions for Constructing Different Kinds of Questions

The following specific suggestions may be of help to teachers as they prepare the different kinds of test questions.

### TRUE AND FALSE QUESTIONS

1. *Avoid specific determiners.*—Words such as only, none, and never usually indicate a statement which is false. Words such as some, should, may, and generally usually indicate a statement which is true.

2. *Present one idea per question.*—Limiting a question to one idea helps avoid complex and ambiguous statements.

3. *Eliminate "trick" questions.*—If the purpose of the test is to measure achievement in the unit of instruction, catch words and double negatives should not be used. Trick questions cause the student to concentrate on how questions are phrased rather than on the meaning of the question in terms of subject matter.

4. *Use short, precise statements of about equal length.*—If a question becomes too involved it should be discarded. Avoid making one kind of statement consistently longer than the other.

5. *Use approximately equal numbers of true and false statements and a random distribution.*—If a pattern of true and false responses is apparent, the students will detect it and adjust responses to it.

The teacher should keep in mind that a true and false test offers the student an opportunity to guess and that it does not test ability to organize. The true and false test is easy to construct, it is easy to score, and it is easy to make comprehensive in terms of sampling broadly the subject matter of the course.

## Matching Questions

1. *Provide more response alternatives than items.*—This will reduce guessing and simple elimination types of response.

2. *Include some response alternatives which are to be used more than once.*—This also reduces guessing and simple elimination responses.

3. *Provide more than one plausible response alternative for each item.*—This will require a more careful discrimination by the student.

4. *Limit the number of response alternatives to 10 and, therefore, the number of items to fewer than 10 in any one matching question.*—Long lists are wasteful of student time in locating responses.

The matching test has about the same characteristics as the

true and false test. It does distinguish between the good student and the poor student better than does the true and false test.

## Multiple Choice Questions

1. *Use simple sentences.*—The questions should not be difficult to read and interpret. Sentence length should not reveal the answer.

2. *Provide plausible incorrect choices.*—The test can be made more difficult by increasing the similarity of correct and incorrect choices.

3. *Provide four or five choices.*—Providing more plausible choices reduces guessing. Obviously incorrect choices should not be given simply to provide four or five alternatives.

4. *Include some questions with more than one correct response.*—Having from one to three correct choices forces the student to consider each alternative more carefully. This technique also helps discriminate among the good and poor students.

5. *Prepare the question so that all choices result in grammatically correct sentences.*—Grammatically incorrect choices usually indicate wrong answers.

The multiple choice test is more difficult to construct than are the true and false and matching tests. In other respects, the three tests are quite similar.

## Completion Questions

The completion type of test questions are statements with a part of each statement deleted and a blank substituted. The part deleted contains the significant information the teacher wishes the student to remember. The statements should be simple and so constructed that:
1. More than one plausible answer is possible.
2. More than one plausible answer results in a grammatically correct sentence.
3. The context of the sentence does not reveal the answer.

Completion tests measure recall more than do the true-false, matching, and multiple choice tests. They are also difficult to construct so that only one answer is correct. Completion tests do provide for comprehensive sampling of course content, and they are easy to score.

### Essay Questions

1. *Prepare each question so that it is grammatically correct.*—Grammatically incorrect statements may cause misinterpretations of the meaning of the question.
2. *Provide specific, clear, and complete directions.*—Lack of clarity can result in many incorrect interpretations of the question.
3. *Provide all the information the student needs to answer the question.*—Missing information results in both confusion and loss of time.

Essay tests limit the area of subject matter which can be covered. They are also difficult and time-consuming to grade. Essay tests are easy to construct, and they measure recall and the ability to organize material.

### Problem-Solving Questions

1. *Provide all needed information.*—Missing information results in confusion and a loss of time.
2. *Give clear and concise directions.*—If the student is to provide some of the needed information from memory, this should be clearly stated in the directions.
3. *Indicate the desired form of the answer.*—Problems can often be solved and answered in more than one way. If the answer should be in a particular form, this should be stated.
4. *Indicate the importance of method of attack.*—If credit is to be given for the correct method of attack on the problem, the student should be told to show all of the work.
5. *Provide all needed materials.*—Rulers, squares, or other equipment should be on hand.

Depending on the nature of the problem, the problem test can sample subject matter broadly, it can measure ability to organize, and it can test for recall of facts. Problem questions are quite difficult to construct, since so much detail must often be supplied, but scoring is relatively easy. Many teachers confuse the essay question, in which a topic is discussed, with the problem-solving type of question.

## TEACHING RESOURCES

A list of the references, teaching techniques, and teaching aids to be used in connection with the study of a particular unit or problem area is an important part of any teaching plan. Many teachers seem to find the lists most useful when they are prepared for each problem area.

In addition to the steps to be taken in solving the individual problems, the teacher needs also to decide on the specific teaching techniques, aids, and references to be used with each problem. For some areas of instruction, the same resources may be appropriate for all of the problems listed. For other areas of instruction, a different resource may be required for each problem. Usually, the final list of resources will result in some variety but not to the extreme of a different technique, aid, or resource for each problem. Field trips, demonstrations, lectures, discussion, student panels, supervised study, movies, visiting speakers, role playing, and television are some of the techniques which can be used. In each case, plans for using a particular technique should be made well in advance, especially for field trips, demonstrations, and the use of visiting speakers. Listing the resources as part of the teaching plan will serve to remind the teacher of the arrangements or materials which must be made or secured. For a detailed discussion of the process of selecting teaching techniques and aids, see Chapter 4.

### Daily Planning

The emphasis on planning in this publication is on a plan which may last from two days to more than three weeks. It should not be concluded, however, that there is no place for daily planning.

As was indicated in the beginning of this chapter, the teacher

# UNDERSTANDING PLANNING 29

needs to keep a record of the problems raised by the group and the progress made in solving them. The teacher may also need to think through and write down, as a part of the daily plan, questions relating to specific problems to help in leading the discussions on those problems. In addition to this, there is a need for planning of a daily nature for many of the special events and activities which are listed in the plan.

One such activity which would require some additional planning is the demonstration. At the very least, the teacher must think through and outline the steps to be taken in order to make sure of having all the necessary materials and that the physical arrangements will be satisfactory.

Another activity requiring additional plans in the form of a daily plan is the field trip. If field trips are to be successful when evaluated in terms of their educational value, it is necessary that they be well planned—including a trial run by the teacher to

THE TEACHER NEEDS TO PROVIDE LEADERSHIP SO THAT
STUDENTS WILL KNOW WHAT TO LOOK FOR.

check on such items as transportation, timing, arrangements with all personnel involved, the order in which various points will be taken up or observed, and the effect of the physical environment. To do otherwise would probably result in a "holiday excursion" rather than an educational experience.

The use of movies, filmstrips, and other teaching aids also requires pre-planning. Previewing by the teacher to determine the contribution to be made by various teaching aids would result in the elimination of many of the aids with a more meaningful use of those retained. It is also very important to arrange in advance for proper seating and to check on the operation of all equipment required.

Daily planning for these and similar activities is necessary and will require greater detail for beginning teachers and for those teachers who have been teaching for a long time but who are experiencing this type of activity for "the first time." The teachers with complete programs will have no time to plan in detail for every minute of every day. It would be better for them to spend their planning time on those plans which will help them "think on their feet."

**Daily Plan Sheet**

The teacher should always have plans ready for the next problem area or other segment to use at whatever point in time the instruction on the current topic is completed.

Following is a list of the headings in a typical daily plan with some additional comments on the purpose of each:

| Heading | Purpose |
| --- | --- |
| Class | To identify the class for which the plans have been developed. |
| Date and Period | To identify plans with the day and time of day for any needed chronological checking. |
| Roll Call | To assure accurate attendance records. Reference can be made at this time to contacts the teacher has had with parents, to recognitions earned by students, or to other personal items of information about students which demonstrate the teacher's interest in them. |

| Heading | Purpose |
|---|---|
| Arrangements | To make sure that everything needed is available and in place. The students should be informed as to what materials and notebooks they should bring to class. The teacher should check on seating, ventilation, and lighting and make any necessary adjustments in plans because of student absences or other late developments. Preparation for demonstrations is especially important. |
| Announcements | To provide for administrative type information releases. School program announcements, FFA meeting dates, and matters of general interest can all be brought to the attention of the class. |
| Review of Previous Day's Instruction | To link the instruction of the previous day's class with that of the current class. The linkage should include a brief statement on the topic of instruction and at which point the instruction ended the day before. Special problems in dealing with classwork or lab activity should be noted. The problems or incidents may relate to safety practices, to behavior problems, to difficulty with a task, or to some other happening. |
| Starting Place for Current Day's Instruction and Order of Events for Continuation | To provide a sense of order and control of instruction. The daily plan should indicate the specific problem being studied and where in the problem-solving steps the class ended. The rest of the problems, or other activities, should be listed in the exact sequence in which they will be handled. |
| Ending the Class | To end the class on a positive note and to make sure that the students are aware of the accomplishments for the day by offering a brief review of the progress made toward accomplishing the student objectives and the teaching / learning objectives. Assignments and some indication of plans for the next day should |

| Heading | Purpose |
|---|---|
| | be given. Finally, student notebooks should be returned to the shelves; tools and materials used should be replaced; and the classroom or lab made ready for the next group of students. The class should end on time so that students will not be late for their next class. |

# CHAPTER 2

# ▶ Theory and Practice ◀

THE EMPHASIS ON "TEACHING FOR UNDERSTANDING" is undoubtedly the most important single development in the improvement of teaching in recent years. Whether the terminology used is *exploration, discovery learning,* or some other term, the emphasis is still on the development of a teaching / learning process which will result in student understanding of concepts and processes rather than on rote learning. In effect, educators are attempting, to a greater extent than ever before, to put into practice the knowledge about learning which has been accumulated over the years.

The problem-solving approach in teaching described in this publication makes greater use of the accumulated knowledge about learning than does any other approach to teaching. Problem solving is the basic process of exploration and discovery. A brief review of what is known about learning is essential to a full understanding and appreciation of a true "problem-solving" approach in teaching.

## SOME PRINCIPLES OF LEARNING

A review of the literature will point out that many principles of learning have been stated and that all of them help to describe some of the basic concepts of learning. Three particular principles of learning which have a direct relationship for the effective use of problem solving in teaching are: the principle of practice, the principle of effect, and the principle of association.

### Principle of Practice

*The principle of practice is that a person learns what is practiced, and continued practice or use is necessary for retaining what has been learned.*

The teacher must not accept the concept that the principle of practice infers a mere repetition of an act. Mere repetition is an unproductive way to approach teaching and will not encourage quality learning. A person could repeat a skill, act, or learning over and over and still be doing it wrong. Thus, a quality control attitude must be held by the teacher to assure that the students are perfecting the skill as it should be performed. Learning also implies that a change in behavior occurs. Thus, as a student continues to practice, the learning taking place should take on more of a quality effort and automatic response on the part of the student.

MERE REPETITION IS AN UNPRODUCTIVE
WAY TO TEACH

The principle of practice may at first glance seem to apply only to the psychomotor or manipulative skills being taught to students. A reflective attitude on the part of the teacher will soon highlight the fact than this principle of learning applies equally well to the development of affective and cognitive learnings. For example, the mental capability of a student will be expanded if the student is continually challenged to use mental abilities to solve problems. The modern-day pocket calculator can serve as an excellent model of how not to pattern learning activities. Students being taught mathematical concepts and the ability to solve prob-

## THEORY AND PRACTICE

lems through the use of a pocket calculator will soon find that they have difficulty solving problems without a calculator. Their mental capabilities will not have been put to use. Likewise, the development of desirable affective abilities requires practice and continued practice to develop perfection, or at least an acceptable level of behavior.

Figure 2-1 illustrates the importance of the need for students to practice what they have learned. When a teacher starts to teach a subject or skill, it is assumed that the student's knowledge of that subject or skill at that point is zero. As the teacher guides the learning process, the student will eventually reach the "threshold of reaction," or that point at which the student has mastered the subject or skill. If the teacher stops instruction on this particular topic at this time, or does not provide the opportunity for the student to practice what has been mastered, the student's ability to repeat the learning successfully will decrease as time goes on (line B). However, if the teacher plans for meaningful practice by the student on that particular topic, then the learning will be retained at a higher level as time goes on (lines C and D). The question the teacher faces once the student reaches the threshold of reaction is, "How long should students be permitted to practice the learning just mastered?" Unfortunately, there is no good guideline to suggest that can be used in different teaching situations and for different students. Teachers cannot afford to devote the amount of

**FIGURE 2-1**

**Relationship of Practice to Learning[1]**

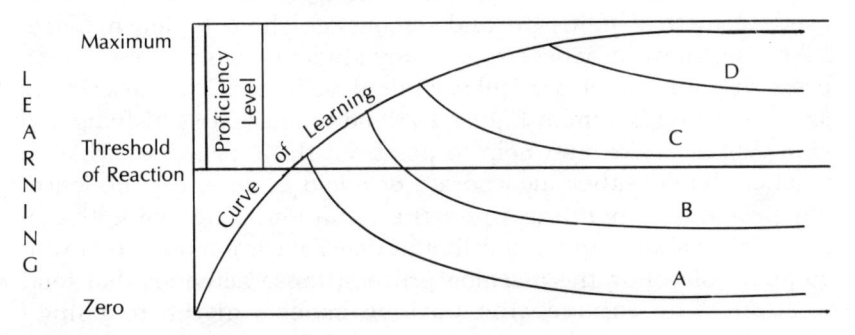

---

[1]Adapted from Arthur I. Gates *et al. Educational Psychology.* New York: The Macmillan Company, 1950.

class time needed to reach line D in Figure 2-1, for if they did, the number of topics covered would be reduced. When to move on to the next topic of learning must be a subjective decision by the teacher based upon factors existing in the local program, that is, the nature of the topic that follows, time available, and student readiness for the next topic. Safe work habits are examples of practices the teacher would want the students to master before moving a class into the laboratory. Students could injure themselves or endanger the lives of fellow students and the teacher if they did not know how to use safe work practices with the various power tools in the laboratory.

The agricultural teacher has a tremendous opportunity to apply the principle of practice. FFA organizations, supervised occupational experience programs, mechanics laboratories, school farms, and land laboratories are just a few of the means by which a teacher can offer students the opportunity to practice what they have learned and encourage them to perfect their skills.

**Principle of Effect**

*The principle of effect implies that if students are satisfied with a learning activity, it tends to promote further student learning and encourages further practice on the part of the student.* Conversely, if students are dissatisfied or annoyed with a learning activity, it tends to hinder student learning and discourages further practice on the part of the student.

A teacher can benefit by using this principle of learning either way as described. Those practices that please students can be used to further motivate and encourage them to learn more. Likewise, those practices that annoy students can be used to express disapproval of particular student behavior. The descriptors listed on a continuum in Figure 2-2 tend to illustrate satisfying and annoying activities and help to point out those practices which a teacher should either incorporate or avoid in directing the learning process, depending upon the situation. The point that a teacher must keep in mind is that as decisions are made on how to go about directing the learning process, those activities that tend to support or enhance the positive aspects of the teaching / learning process ought to be practiced. Likewise, those activities that will tend to be a negative influence on the teaching / learning process should be avoided.

## FIGURE 2-2
### Satisfying and Annoying Aspects of the Teaching / Learning Process

| **Positive** | **Negative** |
|---|---|
| Approval | Disapproval |
| Recognition | Neglect |
| Success | Failure |
| Activity | Inactivity |
| Ownership | Non-ownership |
| Confidence | Fear |
| Creativity | Dullness |
| Excellence | Incompetence |
| Service to Others | Self-centeredness |
| Security | Insecurity |
| Freedom | Restraint |
| Zest | Boredom |

# TEACHING AGRICULTURE THROUGH PROBLEM SOLVING

INACTIVITY IS ANNOYING

## Principle of Association

*The principle of association implies that those experiences that first occur together tend to recur together.* The experiences mentioned in the definition could be abilities, attitudes, skills, or knowledges. This principle could best be illustrated and remembered by all of us when we learned our ABCs.

The implication of this principle of learning encourages the

TEACHING TASKS AS THEY OCCUR IN
REAL LIFE ENHANCES LEARNING

teacher to arrange the teaching / learning process so that desired experiences do occur together. Furthermore, in teaching agricultural subjects, events and activities often occur in a sequence. The teacher should teach this agricultural content in the sequence it actually follows in real life to increase student learning and retention of what has been learned. As an example, the successful transplanting of a shrub should follow a set procedure. If this approved procedure is taught to the students in the recommended sequence, they will be better able to learn and retain the recommended procedure for transplanting shrubs.

The combination of teaching theory and practice is also enhanced by the principle of association. A teacher should never teach theory without the inclusion of practice and vice versa. The effective combination of book theory and practice by the teacher makes for a more productive teaching / learning situation.

## SOME CONCEPTS ABOUT LEARNING

The concepts about learning which follow are illustrative of those which need to be considered carefully in planning for teaching. They form the basis for any sound teaching / learning process. To the extent that any approach to teaching ignores or violates these concepts about learning, to that extent the approach to teaching is a failure. (There is no significance to the order in which the concepts are presented.)

1. *The greatest degree of understanding is developed when the individual sees the parts of the problem situation in relation to the whole problem situation.*—Psychologists theorize in terms of meaningful wholes and going from the whole to the parts. This concept is involved in the identification of enterprises and problem areas. It is also involved in the decision to develop the entire problem area with the class before proceeding to the study and solving of any of the problems developed by the class.

2. *What is learned is most apt to be remembered if it is learned just before it is to be used by the learner.*—That which is learned but not used in some way is soon forgotten. A close identification of instruction with supervised agricultural experience programs and the definition of problem areas in terms of seasonal activities provide possibilities for early

application of learnings that are understood and appreciated by students.

3. *People are "goal seeking" creatures, and they learn best when they have goals for which to strive.*—The use and development of objectives are another part of the teaching / learning process which is often not understood and which, therefore, is avoided by many teachers. Students need to have their purposes or objectives clearly in mind if they are to make progress toward them.

This concept should receive broad applications in teaching in order to make the seeking of solutions to problems a personal challenge to every student. This will help teachers recognize when they have fallen into the trap of trying to substitute their own objectives for the objectives of the students.

The concept also applies to the teacher developing teaching plans. The teacher is also seeking to accomplish certain objectives. These objectives must be identified and defined if the teacher is to know what behavioral changes are to be brought about in students and be able to evaluate teaching in terms of the changes which actually take place.

4. *The teaching of subject matter is a tool to be used to contribute to the development of the individual student as well as a means of helping students acquire knowledge in a particular area.*—The conventional teaching of subjects for the sake of the subject only does little to advance the growth of healthy personalities. The use of a problem-solving approach in teaching places primary attention on the learner and on the needs of the learner, thus contributing to the development of a healthy personality. Problem solving makes the acquisition of knowledge a necessary tool in the achievement of an objective rather than an end in itself.

5. *Concepts or principles are capable of transfer to new situations more readily than are facts.*—Teaching a student only "how to do" a particular job provides little help in dealing with situations which are similar but which necessitate changes in application of facts or which require the gathering of additional facts. It is just as important in vocational education as it is in any other subject matter area to de-

velop understandings of the *why* so that transfer of learning to other situations will be facilitated.

6. *Learnings are most likely to be remembered if learned in relation to situations similar to those in which they are later to be used.*—The more nearly the learning situation approximates that in which the learning is to be used, the easier it will be for the individual to make the transfer of learning from the "learning situation" to the "use situation." Youth often fail to recognize the value of what they have learned in school simply because there has been no understanding developed of the relationship between the learning situation and the use situation. This kind of understanding is most easily developed by using demonstrations and field trips to show the student the application of learnings in actual on-the-job situations.

7. *That behavior which the individual finds most satisfying in a particular situation is most likely to recur when the same or a similar situation occurs.*—If a particular kind of behavior brings success or reward in a given situation, it only makes good sense to repeat the behavior when again confronted by that situation. This principle is basic to nearly all of human behavior. Good teaching attempts to provide for the continuous experiencing of success by the learner. The reward or feeling of satisfaction can take many forms. It can be a simple "very good" from the teacher or the opportunity to engage in new and different kinds of activity. The most important reward is the personal feeling of satisfaction that comes from having achieved. Continuous failure leads to discouragement and ceasing to try. A problem-solving approach in teaching provides many opportunities to encourage students through appropriate rewards.

8. *The individual is encouraged to continue striving by being able to see progress.*—Evaluation of the quality of a performance provides a measure of success, but this alone is not enough. It is also important that the individual be able to see progress along the way. Taking one or two problems at a time in the problem-solving process makes it possible for the student to see another form of progress as the list of problems is slowly but surely completed.

9. *Providing for a variety of activities in the learning situation helps maintain a high level of interest on the part of the students.*—Even the brightest students may tire of studying a single topic or of performing one kind of activity if either is continued for too long a period of time. The teacher must use judgment in planning for teaching in relation to the activities to be used and the breadth of the topic to be studied. This is, in part, the reason for the need to define carefully both units and problem areas.

10. *The experiencing of success or reward must be clearly associated with the desired behavior in the mind of the learner if it is to be most effective in stimulating the desired learning.*—Rewards too far removed from the desired learning or behavior may not be connected with the desired behavior by the learner and will not be effective in stimulating the desired learning. It is also possible for delayed rewards to be connected with some other behavior of an undesirable nature which would then tend to be remembered by the learner. The use of discussion in problem solving provides an excellent opportunity for the teacher to reward student effort immediately after the effort has been made. The reward can also be tailored to fit the individual need.

11. *Practice, to be effective, must be accompanied by continuing evaluation to assure practice of desirable behavior and to provide information about improvement as a reward for the effort expended.*—It is quite possible to develop undesirable behaviors as a result of practice without evaluation. Without evaluation, it is quite possible to practice unknowingly the wrong kind of behavior or fail to make any improvement. Knowledge of success also provides the incentive or reward needed to motivate continuing practice. The step by step approach in problem solving with constant checking against objectives helps prevent the practicing of errors.

12. *Threats and punishment are not useful teaching tools.*—Threats and punishment are as likely to create an unfavorable climate for learning as they are to create a favorable climate for learning. Threats and punishment may actually

inhibit further learning and may make it difficult for the teacher to maintain the kind of relationship with the learner which is needed for maximum success in teaching.

13. *A "readiness to learn" on the part of the learner is necessary for successful teaching.*—A certain amount of growth and development, both physical and mental, is needed before some things can be learned by any individual. In addition, the individual also needs to recognize that learning something has meaning beyond learning for the sake of learning. Part of the readiness to learn is a product of growth and maturation; part of it is a result of such activities of the teacher as helping students set goals and relate learning to accomplishing the goals set.

14. *Individual students will strive harder to achieve those goals or objectives which they have set or accepted as their own.*—The teacher needs to help the learner recognize the true extent of the learner's ability. Much of the success in teaching comes from helping the learner to set goals or objectives which, although achievement is possible, are a real challenge to the learner. Goals which are too easily accomplished do not stimulate the learner to make a maximum effort. Goals for which achievement is not possible may eventually kill the desire to try.

15. *Learner participation in planning learning activities will increase the likelihood of success in teaching.*—Both the young and the old like to share in decisions regarding what they will do. An individual who has helped plan a particular activity will be interested because of having helped plan it. Giving the learner as much responsibility as possible also helps develop leadership abilities. Problem solving as described in this publication provides for a maximum of student participation.

16. *Learning is greatest in classroom / laboratory situations where management is based on learner knowledge of the kind of behavior desired and the reasons for it.*—Teachers can be more relaxed in their classroom / laboratory control measures when learners have sufficient knowledge of the kind of behavior desired, and the reasons for it, to do much of their own disciplining. A learner free from the fear as-

sociated with strict, teacher-administered discipline can devote full energies to creative learning activities. Rules for classroom / laboratory behavior can also be determined through a problem-solving approach. See Chapter 11.

17. *Students who do not find success in learning activities will try to find success in other kinds of activity, whether the other kinds are desirable or undesirable in nature.*—Finding or achieving success at something is necessary to the individual. If continuing failure is the only result of efforts to learn, then learning activities give way to activities in which the chances of success are more likely. A maximum adjustment of teaching to meet individual differences is possible through a problem-solving approach.

18. *Individuals who experience only failures eventually cease to act purposefully or rationally.*—Individuals who experience no success eventually appear to "fall apart." They lose any sense of personal worth and cease trying to perform constructive activities. In despair and anger, they turn against themselves and society.

19. *The ability to think is developed by challenging students with problems which are of interest to them.*—The development of the ability to think is not limited to any one kind of subject matter. Students learn to think by trying to solve problems which are of interest to them regardless of the subject matter involved.

20. *The understanding of a concept or principle is developed by presenting or using it in a variety of situations.*—The application of the same concept or principle to several problems helps the student develop an understanding of the principle in the abstract. Application of a concept or principle in only one situation or in connection with only a single problem may result in the learner seeing it simply as a means to the solution of the one problem to which application was limited. The problem-solving approach described here provides for variety in the application of a principle through the different problem areas as well as through the use of a variety of teaching techniques.

21. *The recall of ideas and facts is aided more by answering questions which require the use of the ideas and facts than*

**THEORY AND PRACTICE** 47

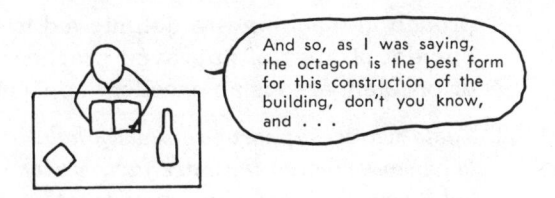

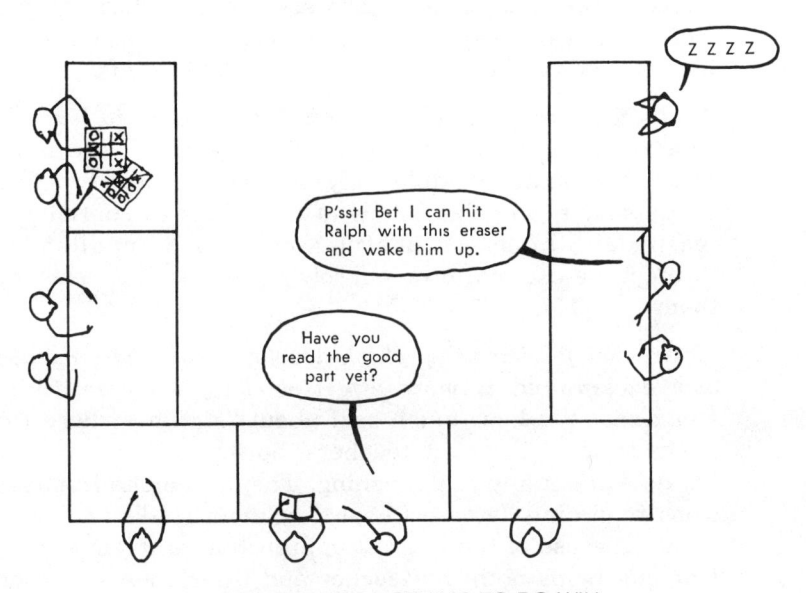

> And so, as I was saying, the octagon is the best form for this construction of the building, don't you know, and . . .

> Z Z Z Z

> P'sst! Bet I can hit Ralph with this eraser and wake him up.

> Have you read the good part yet?

STUDENTS WITH NOTHING TO DO WILL
SELDOM DO NOTHING

*it is by simply reading the same materials over and over again.*—To be most effective in committing information to memory, the information should be recalled shortly after it has been learned. This helps in two ways: First, it helps the memory during the period of time when forgetting is most rapid and second, it helps the student determine those facts which are the most difficult to remember and upon which the student needs to spend more time. Most people need also to review things at intervals in order to commit them to memory for a prolonged period of time. The emphasis on "use" of facts in a problem-solving approach in teaching is a definite aid to learning. The development of a list of approved practices at the end of a unit or problem area is an excellent application of this principle.

22. *Facts which support previously held beliefs are more likely to be remembered than are facts which contradict previously held beliefs.*—Another way to state this principle is the commonly used expression, "We believe what we want to believe." The sharing of experiences which enables students to see that certain tasks are being performed in a variety of ways helps them be objective and accept new ideas which run counter to previously held beliefs.

23. *Adults who continue to learn retain the ability to learn.*— The ability to learn increases with age up to the adult years. The ability of adults may decrease because of a lack of motivation and the failure of the adults to continue to learn new things. Adults also need to be challenged through the use of a problem-solving approach in teaching them.

24. *Motivation for learning which develops from the student's own background is more effective than is an artificial stimulus.*—We hear much said about "starting where the students are." To many teachers, however, these are just words without any real meaning. They still make the decisions regarding "where to start" without student involvement. The use of the interest approach in problem-solving teaching helps both the teacher and the students gain an understanding of what each person knows about a particular area of knowledge and thus helps all to understand and accept the problems of the rest in learning about the area.

25. *The student expects the teacher to be an active participant in the teaching / learning process.*—Students expect teachers to provide leadership in the teaching / learning situation. When the teacher fails to participate, the student loses time in getting started and, in fact, may never really become properly oriented to the learning situation. It is the teacher's task to provide the amount of direction needed to enable the student to make the most effective use of time and to help the student develop the ability to direct the study of any given subject matter area.

26. *Motivation should be developed by helping the learner establish goals which are outcomes of learning rather than by offering prizes or other artificial stimuli.*—The development of a desire for continuing education is more probable if goals or objectives are developed which can be achieved only through learning. Individuals motivated only by prizes and other forms of awards may fail to develop the understanding of the value of learning which is essential to self-enforced participation in learning on a continuing basis throughout life.

27. *Learning helps to make one person different from another.*—The person who learns to weld is different from the person who has not learned to weld. The person who learns to speak well before a large audience is different from the person who has not learned to speak before a large audience. In a sense, individuals become what they are as a result of what they learn. Learning helps create the differences which exist among individuals.

28. *Teaching and learning are of value to the extent that it is important that the person being taught learns what is being taught.*—The significance of the work of the teacher and the meaningfulness of learning are dependent on the potential value of that which is taught to the learner. In essence, the statement made indicates the need for the teacher to have clearly defined objectives before teaching and for the learner to feel that what is to be learned is important.

29. *How well a person is taught influences that person's learning.*—A person who has been taught well has a greater opportunity to develop abilities and skills to a high degree of proficiency and to acquire more of the meaningful and useful kinds of learning than the person who has not been taught well.

30. *The teaching / learning process must be sufficiently flexible to make possible adaptation to individual differences.*—Individuals differ regarding physical maturation, mental development, background, and many other factors. These differences must be perceived by the teacher and adjustments made in teaching for greatest effectiveness.

31. *Some things cannot be learned until other learnings have been acquired.*—Without appropriate prior learning there are some things a person cannot learn. For example, if a task requires the use of certain tools, learning the performance of that task is not possible without a knowledge of the tools to be used.

32. *How a teacher teaches depends, in part, on how the teacher views learning as taking place.*—The teacher who views learning as the memorization of facts need only present these facts to the student and test for them. The teacher who views learning as involving application and understanding must find ways to develop and test for these objectives.

33. *There are many factors which affect the ability to learn.*—Scholastic aptitude is only one of many factors affecting the ability to learn. Physical coordination, fear created by the inability to respond properly, health, social development, speaking ability, environment, and many other factors are also important. The teacher must learn to perceive and deal with all of these many factors as they appear in individuals and groups being taught.

34. *How well a person will learn is best predicted from how well that person has learned.*—The best predictions of future performance of an individual are based on the past performances of that same individual in similar tasks and situations. This should not blind the teacher to the recognition of any efforts at removal of some of the factors which limit the ability to learn.

35. *Something learned is remembered better if frequently used, at least to the point of a high degree of "overlearning."*—A person soon forgets something unless it is used frequently. This use can be at intervals of time sufficiently close to make forgetting impossible, or it can be a use so frequent and intense for a short period of time that immediate recall is possible even with some passage of time.

36. *Learning of some kind, whether desirable or undesirable, takes place in nearly every conceivable kind of situation with or without someone to do the teaching.*—Students who

**THEORY AND PRACTICE**                                              51

whisper to other students while the teacher is speaking are learning to ignore the teacher and to whisper. The person alone in the woods watching birds is learning about birds. The young boy who watches automobile drivers go through intersections without stopping for the stop signs is learning to ignore traffic signals.

37. *The difference in the ability to learn between a slow learner and a fast learner increases as students progress through school.*—Since learning provides the learner with additional tools to use in learning, the fast learner in acquiring these tools increases the ability to learn at a faster rate, while the slow learner acquires the additional tools for learning but will apply these at a slower rate. The teacher who expects the slow learner to "catch up" is simply misjudging the situation. Slow learners actually have a double handicap; not only will they learn more slowly but also they will probably forget more quickly.

38. *The extent of teacher-direction and teacher-explanation provided to the learner should be proportional to the ability of the learner to direct the learning process.*—One of the objectives of the teacher should be to develop the ability of the individual for independent study. The teacher must operate at a level of teacher participation which prevents student floundering and failure but which, at the same time, challenges the student to do just a little more than has ever been done before. The person who is able to think through a situation and develop the explanation learns more than the person who is given an explanation of the situation.

39. *Creative ability is not limited to persons with a high scholastic aptitude.*—Slow learners, as well as fast learners, have creative ability. Teachers need to learn how to recognize and develop the creative abilities of their students.

40. *Individuals learn through their own activities.*—Learning cannot be given to a person; individuals must learn for themselves. Anything which limits learning activities limits learning. Teachers and others can only help.

41. *The evidence that learning has taken place is a change in*

*behavior on the part of the learner.*—The evidence of learning may be as obvious as the ability to perform, and the actual performance of, a physical skill. It may also be difficult to detect, such as a change in attitude or the way a person solves a problem. In either case, the individual has changed in behavior. Different techniques are needed to measure the various kinds of change.

42. *The more vivid a particular learning experience is, the greater is the probability that what is learned will be remembered.*—A vivid presentation tends to capture a person's attention to a very high degree. Since a high degree of attention is associated with efficiency of learning, the more vivid experience will be remembered better than the less vivid experience.

43. *The effectiveness of practice is increased when the length and the spacing of the practice sessions are adjusted to the nature of the task and to the motivation of the learner.*—Practice is usually ineffective if the learner is not motivated to practice desired behaviors. Tasks should not be practiced to the point where physical or mental fatigue sets in and interferes with the performance of the task.

44. *Transfer of learning is more likely to take place if the teacher "teaches for it."*—The teacher who has as a teaching objective the development of the student's ability to transfer learning to new situations will include activities in teaching which will bring about this result.

45. *Something becomes "meaningful" to a person as it is experienced.*—A person's perception of a particular thing is made up of personal experiences with it. Persons see objects, ideas, or themselves as they have learned to do so. The sheep, seen from a distance, is still perceived as a timid creature covered with soft, thick wool. Individuals may perceive themselves as being unable to perform well under pressure because only failure has been experienced in similar situations.

46. *Self-evaluation of learning by a person is related to goals or level of expectation.*—Success for one person might represent failure for another because of different levels of ex-

pectation. The person who has grown only 600 Christmas trees per acre would feel very successful if 900 trees were grown. The person who had been producing 1,200 trees per acre would consider a yield of 900 trees per acre as quite unsatisfactory.

47. *People like to do what they do well and avoid doing what they do not do well.*—Students must be helped to recognize the importance of doing things which they do not do well, and they must then be provided with the opportunity to do them. Too often, especially in extra-class activities, the student who is already a good leader gets additional leadership experience while those who badly need this experience receive none.

48. *Knowledge can be memorized and used without understanding having been achieved.*—A person may know about a herbicide and use it correctly to kill a particular weed without understanding how the herbicide acts, why it is applied in a particular way, or why it does not also kill the corn plant. Meaningful learning, learning which can be applied to new situations, depends on understanding the *how* and the *why*.

49. *A problem cannot be solved until it has been identified and clarified.*—Much energy and a great deal of time are wasted in aimless activity by students and others simply because the problem on which they are supposed to be working has not been well defined. A poorly defined problem does not serve to provide direction to the learner as will a well defined problem.

50. *Learning of skills is made easier by seeing the results or product of a skill prior to seeing it demonstrated.*—It is easier to see and understand the demonstration of a skill if there is some knowledge of what the demonstrator is trying to accomplish.

51. *Skills should be learned as integrated wholes, not as isolated parts to be assembled later.*—A skill divided into separate acts and taught as a series of unrelated parts may never be well performed by the learner. The skill consists of something more than the sum of the parts This does not pre-

clude the teaching of a skill step by step with each step in proper sequence.

52. *Both speed and accuracy must be considered in teaching skills.*—Ordinarily, accuracy in performing a skill is stressed first under the assumption that speed will automatically follow. However, some skills must be performed rapidly from the start in order to obtain the desired performance.

53. *Teaching is most effective if the learner respects and likes the teacher.*—Respect is more important than liking, but a teacher who has both the respect and the liking of the students can expect them to work harder to learn than would be the case without this respect and liking. Both teacher and student energies are wasted in dealing with problems growing from a lack of respect and liking.

54. *People learn more effectively through observation if they have something to look for.*—It is much more probable that the learner will see the most important things in a movie or other observation if some direction is provided regarding what to look for.

55. *Learning contributes to habit formation through influencing what is done.*—Knowledge of something does not develop a habit. A habit is developed by doing something many times in the same way. Knowing how to hold a tool does not mean that the learner will hold it properly when using it.

56. *A habit is replaced or broken by learning or by providing something to take its place, something which is more satisfying, or, at least, less dissatisfying.*—Teachers should remember, if they wish to change a student's habit pattern, that some other form of behavior must be substituted for the behavior to be changed or eliminated.

57. *People learn from each other.*—People are social by nature and enjoy associating with each other. To eliminate all group instruction with its potential for having students learn from each other would indeed be poor judgment.

58. *A person's attitudes toward something are major determiners of what that person will do in relation to it.*—The

development of favorable attitudes is also an objective of education. Motivation is in part the development of favorable attitudes toward doing or learning something.

59. *The nature or kind of instruments and processes used by the teacher to evaluate achievement in learning influences the structure of the learning process developed by the learner to learn.*—Able students soon learn the difference in preparation needed to do well on true and false examinations as opposed to problem-type examinations. They also recognize the difference in scope of content which can be included in each kind of examination and govern their study habits accordingly.

## SOME RELATED STUDIES

There are many studies which support the concepts and principles stated. A few of these studies are reviewed in the following pages to illustrate what is available.

One of the values inherent in a problem-solving approach in teaching is that it is not a process which is strange or unused by people in general. It is used in one variation or another in solving a great variety of problems. Some support for this point of view can be found in a psychological study conducted by Durkin.[2] She had a two-fold purpose: (1) to observe the processes and patterns involved in a single form of the problem-solving type of thinking and (2) to test any new hypotheses that the results might suggest, with particular relation to the meaning of the material in light of the controversy over trial and error and insight.

A selected group of 12 college graduates and 16 graduate students in psychology at Columbia University were used in the study which consisted of two parts. In the first part, 21 of the selected group were each given four puzzles to assemble. Each puzzle, when properly assembled, formed a Maltese Cross. For the second part of the study, all the pieces used in the four puzzles for the first part were used in forming a total (large) Maltese Cross. All 28 persons (7 of whom had had no previous experience)

---

[2]Helen Elsie Durkin. "Trial and Error, Gradual Analysis and Sudden Reorganization, An Experimental Study of Problem Solving." *Archives of Psychology*, No. 210, 1936.

WRITING THE PROBLEMS ON THE BOARD IN THE WORDS OF THE STUDENTS (INSOFAR AS THIS IS POSSIBLE) IS VERY IMPORTANT.

participated in the second part. Each person worked individually and was informed as to the purpose of the experiment—that it was not an I.Q. test, that the experimenter was interested only in the method used in solving the problem, and that the person working on the puzzle should talk giving expression to all thoughts and ideas occurring during the solution of the problem. A record was kept of remarks, moves, and errors made as well as time consumed.

Durkin found that some subjects relied almost entirely upon manipulation in solving the problem, with many errors and apparently little perception of relationships. Some worked in groping fashion, stopping suddenly and then proceeding rapidly toward solution; while others worked quite deliberately and with concentrated attention.

It was concluded that problem solving is never completely

## THEORY AND PRACTICE

blind or random, although it is in some degree exploratory and always requires some manipulation. It is not that a new kind of process—that of seeing the relationship of material to goal—enters the level of insightful thinking, but that the process was there in embryo even in the blindest "trial and error."

It was observed that three principal types of solution were distinguishable: (1) trial and error—characterized by blind groping, hindsight, confusion till the last moment, apparent hopeless feeling, attention to a distant and fuzzy goal, wandering attitude and haphazard, irregular error curve, with poor transfer and baffled manner; (2) sudden reorganization—characterized by groping suddenly stopped, sudden foresight, abrupt clearing of confusion, excitement, elation, relief, a looking for wholes and relations, attention not centered on goal, receptiveness, irregular suddenly dropping error curve, good transfer, baffled, suddenly well organized and efficient; and (3) a gradual analysis—characterized by no groping, progressively developed understanding, foresight, step by step clearing, satisfaction, concentration of attention on specific goal needs, step-like error curve, good transfer, calm and well-organized manner. Each stimulus response was preceded by a short, intent pause—the person seeming to try to grasp a series of rapid, fleeting inferences and recalls to bring focus. Stimulus response paralleled description of hidden insight and was found to be related to certain previous responses to the material, to observations, recalls, and inferences that may have been related to earlier experiences during solution or to previous experience. The same processes were found throughout, that is, observation, recall, perception by relationships, and attention to goal. Manipulation is still an additional factor and inference can be considered a higher level of perception. The three forms of thought may be regarded as points on a continuum blending into each other rather than exclusive kinds of thinking.

Additional inferences could undoubtedly be drawn regarding the principal types of solutions distinguished, but Hildreth[3] probably presents a clearer picture in relation to a problem-solving approach to teaching—both with regard to use and with regard to the effect on the pupils.

Hildreth attempted to evaluate the effects of understanding

---

[3]Gertrude Hildreth. "Puzzle Solving With and Without Understanding." *Journal of Educational Psychology*, Vol. 33, pp. 595-604, 1942.

on puzzle solving at a somewhat different level. One hundred children, 60 boys and 40 girls, aged 7 to 10 years, were selected from a private elementary school for the study. To equalize experience, each child practiced with a 7-piece jigsaw puzzle. Then each child was given three different 21-piece picture puzzles to assemble while being observed. On the first of these the children were uninstructed; on the second they were instructed to the extent of being shown the assembled puzzle and having it described by the examiner. Then it was disassembled and given to them to work. The third puzzle was given to them disassembled and upside down so that there were no picture clues to guide them. A record was kept of moves made and time consumed in assembling each puzzle.

It was found that the problem was easier and more quickly done when approached in a meaningful way; that is, results were better when the children worked with understanding and when separate pieces were understood in their relationship to the whole. Working with the puzzle wrong side up was more difficult than working with the picture side up. The problem was easiest when the children had seen the puzzle assembled before trying to work it.

Hildreth concluded that: Solving problems in school will have meaning for students to the extent that they have been given an overall view of the whole, some indication of central meaning, and some clues to solution. If the student has an organized goal in mind toward which effort is or can be directed, it should expedite problem solving. Thus, students should be more efficient at working problems that have meaning for them than trying in a mechanical way to master detached, unorganized fragments that have no meaningful associations.

A third study appears to lend support to the proper use of the "interest approach" as previously described. This study, conducted by Weaver and Madden,[4] was designed to evaluate the effect of direction on problem solving. This was actually a repeat, with some modification, of an earlier study by N. R. F. Maier on reasoning in humans. However, it was a somewhat different problem conducted in a different setting. Each of 84 selected subjects

---

[4] H. E. Weaver and E. H. Madden. "Direction in Problem Solving." *Journal of Psychology*, Vol. 27, pp. 331-345, 1949.

was given appropriate materials and asked to construct a device which would suspend two pendulums to swing over two spots on the floor.

The subjects were divided into four groups. Those in one group were given a statement of the problem only; those in a second group were given a statement of the problem, a demonstration of the parts, and a statement indicating that if the ideas given them were combined in the right way, they would have the solution to the problem. Those in the third group were given a statement of the problem and a hint that it would be simple if the pendulums could be hung from a nail in the ceiling. Those in the fourth group were given the statement of the problem, a demonstration of the parts along with a statement on their relevance to the problem, and the hint or direction given to the third group. Twenty-five minutes were allowed for working on the problem, during which time a record was kept of trials and errors made.

The subjects appeared to respond according to previous experiences with the materials. Some were ready to use the correct procedure even before they received directions. Those who solved the problem produced solutions quickly. Several persisted in following the same pattern in a wrong approach to solution. All attempts, even when unsuccessful, reflected a pattern. In all, 10 persons solved the problem—1 who had received directions only, 4 who had received the "parts" explanation only, and 5 who had received the directions plus the "parts" explanation.

The authors indicate that the make-up of things and the arrangement of events, as related to experience, will circumscribe the field in which new discovery may occur. Language and understanding and recall of relevant experience, coupled with skill in self-direction, are important; while social climate, level of aspiration, attitudes, and self-regard operate to keep one in action until an acceptable response relieves tensions. Habits of searching and recombining give direction and new direction to perceptual responses, and changes in perception depend upon unities that have been identified within the perceived field.

The statement has been made throughout this writing that if a problem-solving approach is to be effective, it must be based upon the problems faced by the students in their experience programs. The principle has been violated by even good teachers because of the feeling that the course of study would be too limited. A consideration of a study by Collings[5] may help relieve fears on this score. He conducted an educational experiment dealing with the elementary curriculum to discover whether the country school curriculum could be selected directly from the purposes of boys and girls in real life; and if so, to what extent, with what effect, and under what conditions. The experiment covered the four-year period from 1917 to 1921.

The study was conducted in three rural schools in MacDonald County, Missouri. One, in which 41 children ranging in age from 6 to 15 years were enrolled, was set up as the experimental school and the other two, in which a total of 60 children ranging in age from 6 to 16 years were enrolled, were the control schools. The equivalent group method was used in which the children in the experimental school were paired as nearly as possible with children in the control schools on the basis of intelligence level, chronological age, years of schooling, number of years spent in the schools included in the experiment, school achievements and attitude toward school, socio-economic status, and parentage.

The control schools provided the traditional curriculum, following a syllabus set up by the state for elementary schools. The experimental school provided a project activity curriculum based on the problems and purposes of the children.

The children in the experimental school were divided into three groups, primarily on the basis of age, to facilitate handling from the standpoint of complexity and type of work to be done and to minimize confusion in work conferences. Teachers worked with children in these conferences from 30 minutes to 1½ hours daily, depending upon the extent to which the different groups happened to be working on similar materials or project problems. The curriculum was based on four types of projects—story, hand, play, and excursion projects—which arose out of the problems identified by children.

The excursion projects consisted of visits to homes and businesses in the community to observe and study practices and identify problems. Upon returning to school, the children analyzed problems identified—referring to necessary references and other sources of information—arrived at conclusions, and wrote up recommendations. These recommendations and a complete account

---

[5]Ellsworth Collings. *An Experiment with a Project Curriculum.* New York: The Macmillan Company, 1923.

of the trip were frequently presented to the public at a community meeting.

Usually most problems led to associated projects which, in turn, were set up and solved. For example, one child was absent from school because of illness with typhoid fever. The children knew that this particular child's family had had several recurrences of the disease and that one member of the family had died of typhoid. They decided to make a study of the home situation to try and identify the causes, and possibly solve the problem. They found that sanitation was poor around the home, due to exposed garbage and sewage. They found that there were hordes of flies and that there were no screens on the doors and windows of the house. In a search through various information sources, the children discovered that flies were carriers of typhoid. They recommended the installation of screens and a general clean-up, both of which the family was happy to do. This led to an associated project in the form of a survey of disease and sanitary conditions in the entire community.

In similar fashion, problems in hand projects, such as building an ironing board, proceeded through the various stages of planning, executing the plan, and criticizing the product. These, in turn, led to such associated problems as how an electric iron works and what is included in the price of ironing boards.

Play projects consisted of games, dramatizations, folk dancing, and parties with all the attendant problems of selection, planning, and execution. Story projects consisted of whatever was felt to be needed. Various selections were prepared and presented, including nursery rhymes, stories from library books, song stories, and picture stories.

During the four-year period, careful records were kept on everything that happened, both in the communities and in the schools, which bore any relationship or was traceable to the effect of the school program. Also, national standardized achievement tests for penmanship, written composition, spelling, American history information, geographical information, reading comprehension, and accuracy in addition, multiplication, subtraction, and division were appropriately administered to all students in the three schools.

From the records kept and the results of the achievement tests, it was found that the experimental school students were superior to the students in the control schools in all areas of achievement. Differences ranged from 18 to 71 percent, with the greatest difference in performance being in geography. The experimental school children had developed better attitudes which were reflected in a higher percentage of enrollment, a higher percentage of daily attendance, and decreased tardiness; more children continued into the upper grades; and a higher percentage of graduates went on to high school. There was a greater overall improvement of students in the experimental school accompanied by a decrease in truancy and corporal punishment.

Also, parental attitudes were improved in that from 44 to 85 percent more parents visited school, voted for maximum teacher levy, voted for school improvement, used the school in the solution of farm problems, used the school library, and voted in favor of a rural high school in the community served by the experimental school than was the case in the control schools' area. In addition, the experimental school was influential in getting from 29 to 57 percent more of the homes to start libraries, buy magazines and newspapers, place screens on doors and windows, initiate yard beautification programs, install modern conveniences, and purchase musical instruments.

It has been concluded that the school curriculum can be selected directly and entirely from the real life purposes of boys and girls. The achievement of boys and girls where the problem curriculum was used was superior, which indicates that school work in terms of the purposes of boys and girls is practicable in any American rural school. However, to assure success some

CLASSROOM WORK, LAB WORK, AND EXPERIENCE PROGRAMS SHOULD BE RELATED.

basic requirements in the way of training teachers, providing facilities and libraries, and suggesting use of available money in the purchase of supplies and equipment must be met.

Another study which has direct implications for the source of the content in vocational agriculture is the eight-year study dealing with curriculum problems on the secondary level.

In reporting on the study, Aiken[6] indicates that the experiment grew out of the dissatisfactions expressed and recognition of the need for improvement of the secondary school curriculum as brought out during the annual convention of the Progressive Education Association in October 1930. These educators recognized the worth of, and generally approved, many proposed suggestions for curriculum revision, but were apprehensive lest any drastic departure from the conventional curriculum would cause their students to be rejected by colleges.

Accordingly, a commission was established to seek an agreement with colleges whereby they would agree to accept students trained in a high school curriculum that had been fundamentally reconstructed. Such a plan of co-operation was worked out with the universities and colleges agreeing to accept the graduates of a small number of representative secondary schools without the usual subject matter and unit requirements for college admission for a period of five years, beginning with the class entering college in 1936.

Thirty secondary schools were selected for participation in the experiment. Included were public, private, large, and small schools representing different sections of the United States. The schools began the new program in the fall of 1933, each developing its own plans and deciding what changes should be made in curriculum, organization, and procedure. Two major principles served as guides in the reorganization of the school programs: (1) that the general functioning of the school and methods of teaching employed should be in harmony with what was known then about human learning and growth and (2) that all changes should be aimed toward rediscovery of the chief reasons for the existence of the high school in the United States.

The principal types of curriculum changes introduced by the co-operating schools consisted of introduction of content which

---

[6]Wilford M. Aiken. *The Story of the Eight-Year Study*. New York: Harper and Brothers, 1952.

had greater significance to students and modification of extreme departmentalization into broader fields of study. Some schools integrated their program by drawing upon all subjects and all teachers to contribute to the broader field with the student's choice of vocation becoming the integrating center for that student's curriculum. The most marked departure was the introduction of a core curriculum which dealt with the common concerns of youth.

A follow-up staff was organized and a very careful study was made of graduates of the participating schools who went on to college. To establish a just basis for comparison, each of the graduates of the 30 schools who entered college was most carefully matched with another student in the same college who had taken the usually prescribed high school courses, met the conventional college entrance requirements, and graduated from some school not participating in the study. Students were matched on the basis of sex, age, race, scholastic aptitude scores, home and community background, interests, and probable future. Comparisons were made in 18 areas of performance.

Altogether 1,475 pairs of students were studied: those entering college in 1936 for four years, in 1937 for three years, in 1938 for two years, and in 1939 for one year. It was found that graduates of the 30 experimental schools were superior in all but 2 of the 18 areas of performance, and in these 2 there was no discernible difference. The college follow-up staff stated that it was quite obvious that the 30 schools' graduates, as a group, had done a somewhat better job than the comparison group, whether success was judged by college standards, by the students' contemporaries, or by the individual students themselves.

It was generally conceded that every student should achieve competence in the essential skills of communication, for example, reading, writing, and oral expression; in the use of mathematical concepts and symbols; and that inert subject matter should be weeded out to make way for content that is realistic and pertinent to the problems of youth and modern civilization. Hence, the common recurring concerns of youth should provide content for the curriculum with the functioning and work of the schools contributing in every possible way to the physical, mental, and emotional well being of every student. The curriculum should have but one clear, major purpose, that of enabling all young Americans to utilize their heritage of freedom, to develop an un-

**THEORY AND PRACTICE**

derstanding of the American way of life, and to generate an inspired devotion to human welfare.

In writing on some implications of the study for high schools and colleges, Aiken[7] indicates that they should join in an attempt to develop a more dynamic education by subjecting present programs to most careful scrutiny and questioning everything they have been taking for granted. Carefully planned programs of changes should avoid undue haste or piecemeal effort, and every teacher should be involved in all deliberations prior to reconstruction of the curriculum. Provision should be made for parents to participate in any changes that take place, and students, who also have important contributions to make, should be invited to think about what the school should be and do. The school should undertake to remake itself thoroughly—a most difficult, but thrilling and profitable, experience—eliminating deadening routine and formulating plans for evaluating, recording, and reporting progress and results.

Brownell[8] has done much to clarify thinking regarding problem solving in his writings for the forty-first Yearbook of the National Society for the Study of Education. He indicates that problem solving may be defined so broadly as to be synonymous with learning; that all learning starts with some inadequacy of adjustment, some disturbance of the equilibrium; that problem solving is simply a process whereby the subjects extricate themselves from such situations; and that problems are distinguished from other situations by the peculiar relationship that exists between the learner and a task. He indicates further that a problem situation cannot be assumed to be a problem to all learners since it is influenced by both highly personal and narrowly individualistic factors and social factors—impersonal appraisal. He also states that according to Parker's analysis, attempts to solve puzzles have little educational value in the development of problem-solving ability.

The relationships necessary to the solution of a problem must be well within the understanding of each child and identifiable by the child with reasonable effort. Skill in problem solving is partly a

---

[7] Wilford M. Aiken. "Some Implications of the Eight-Year Study for All High Schools and Colleges." *North Central Association Quarterly*, Vol. 17, pp. 274-280, 1943.

[8] William A. Brownell. "Problem Solving." *The Psychology of Learning*, Chapter XII, Part II, 41st Yearbook, National Society for the Study of Education, pp. 415-443.

matter of technique and partly a matter of understandings and meanings which have been carefully developed through a wide variety of experiences. However, practice in problem solving should not be just repeated experiences in the same situations, but rather it should be solving different problems by the same techniques and applying different techniques to the same problems. The most valuable experience is organized experience. A problem is truly solved only when the learner understands what has been done and why the actions were appropriate, and when the learner is able to summarize solutions clearly and state them verbally. Mistakes are not *corrected* merely by providing the *right* solution, but rather they are corrected by exposing weakness and substituting sounder attack through the development of needed meaning or understanding and the ability to differentiate between the reasonable and the absurd. Only continued experiences in the solving of real problems will produce an inquisitive mind and a problem-solving attitude. Memorizing and cramming will not.

**THE TEACHER SHOULD LEAD THE STUDENTS IN SOLVING PROBLEMS—NOT SOLVE THE PROBLEMS FOR THEM.**

According to Brownell, both the school and the teacher have the responsibility to aid children in formulating the problem clearly, to see that they keep it continuously in mind, and to encourage suggestions for solution by making them analyze the situation, that is, recall pertinent rules and principles. This means that teachers must insist that students evaluate each suggestion and organize the processes of solution, that is, make outlines and diagrams and formulate concise statements of the outcomes of activity.

From the foregoing, it appears that some extremely valuable

work has been done in pointing the way to more effective methods and content for education. Also, it seems that problem solving, as a technique employed in the two curriculum studies reviewed, can be a most successful and challenging approach to a real and vital educational program, one that goes a long way toward providing for readiness for learning, motivation, organization, and transfer of training—elements that most educators agree are crucial in the learning process.

However, the field of education as a whole appears to have been reluctant to accept that challenge, whether because of lethargy, slavish devotion to tradition, or fear of the unknown.

# ◆ CHAPTER ◆

# 3

# ▶ Motivating Students ◀

PERHAPS ONE OF THE MOST IMPORTANT TASKS a teacher faces in the classroom is that of motivating students. Without proper motivation, students may never get off to the right start in class. Motivating a class for an upcoming topic is an important step for successful teaching / learning.

If one were to review the different lesson plan formats available, one would discover that the first opportunity a teacher has for involving students is usually during the introduction, motivation, or interest section. Even though the teacher may already have prepared some behavioral objectives, identified references to be used throughout the period, or described a lesson topic, it is not until the motivation or introduction section of the lesson plan that the students are actively brought into the teaching / learning process.

In an initial discussion of motivation, the difference between two words, "motivation" and "attention," must be clarified. These two terms are not synonymous, and it is important that a teacher has a clear understanding of each. A teacher may have the students' attention but this does not imply that motivation has occurred. Attention implies that the students are looking at the teacher and are not distracted by other things throughout the classroom or laboratory. Motivation, on the other hand, implies movement toward a goal. Furthermore, it implies that students are beginning to direct their energies and their minds to solving problems which have been identified. Thus, if the first few seconds of activity in the classroom are really analyzed, the analysis will show that a teacher will come into the classroom and immediately begin the task of getting the students' attention. The second step is to motivate the students toward some predetermined goal or objective that has been established for that particular class.

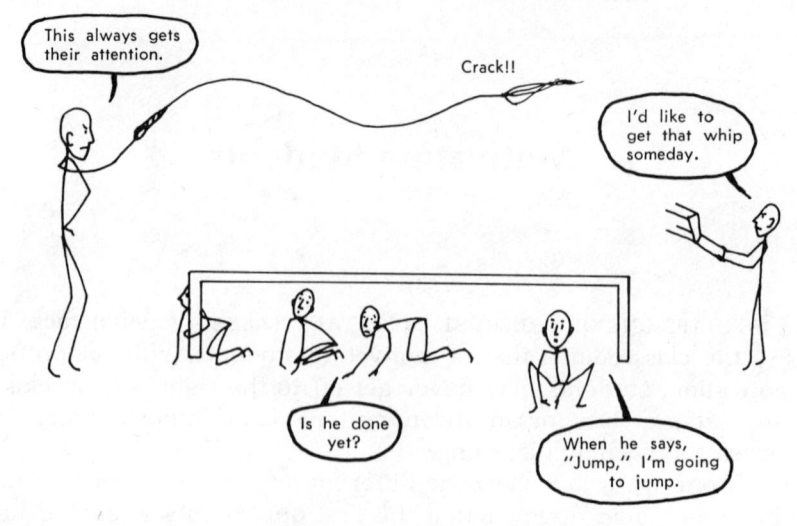

72    TEACHING AGRICULTURE THROUGH PROBLEM SOLVING

IT IS IMPORTANT TO SET A POSITIVE
STAGE FOR CLASS

## ASPECTS OF MOTIVATION

A review of literature will point out that there are two different kinds of motivation, intrinsic and extrinsic. It is essential for a teacher to be able to distinguish between them and to understand the implications of each for motivating students for a particular class. Both will be dealt with in greater depth in the following sections.

### Intrinsic Motivation

Intrinsic motivation implies that the initial action of the student toward a particular goal or reference point comes from within the student. In other words, the student wants to do something and proceeds to set into motion those things that will help to accomplish that particular goal. Each student has some degree of intrinsic motivation, and the challenge to the teacher is to stimulate that interest and motivation to the point at which a student reacts appropriately. However, different students will have varying levels of motivation, and it is important that a teacher become familiar with each student and what it takes to begin the motivational process. Highly motivated individuals have positive pictures

# MOTIVATING STUDENTS

of themselves and are aware of the future. They have realistic appraisals of the past and, overall, appear to be mature. On the other hand, low motivated students tend to have a low self-esteem and feel that they are not liked. Furthermore, they always feel they are unworthy and have nothing to offer to the class. They are sensitive to criticism and tend to have hyperactive attitudes. If left alone, these students tend to be secluded, shy, and timid. It is important that the teacher remember these unique characteristics and adjust strategies in teaching a class when these types of student characteristics surface.

## Fundamentals Which Influence Motivation

There are five basic fundamentals which influence the motivation that students may exhibit. These five fundamentals center around Maslow's[1] theory of learning and each of these five steps will be reviewed.

1. *Need for physiological comfort.*—It is important that these needs are met before any efforts are made to motivate students. These needs include food, shelter, sleep, and water. The fulfillment of these individual needs takes precedence over any other demands which others may want to place on students. Thus, it is essential that the teacher recognize these needs and be able to sense when these particular needs are unmet for any student.

2. *Need for Safety.*—Once the basic physiological needs have been met, the next need deals with safety. Safety needs must be met if other needs of the student are to be fulfilled. Even though all the physiological needs of the student have been met, it will be very difficult for the teacher to motivate that student towards the planned topic if the student is fearful of something or is concerned about personal safety.

3. *Need for love and affection.*—Each individual has worth. Thus it is important for students to realize that they are individuals and that they do have something to contribute to class. In an earlier section of this chapter, it was pointed out that students who have low motivation tend to

---

[1] A. H. Maslow. *Motivation and Personality.* New York: Harper and Row, 1970.

feel that they are not loved. Thus it will be difficult to motivate these students. The teacher would need to make sure that ample opportunity is available to express love and appreciation for these students as human beings.

4. *Need for self-esteem.*—This basic need centers around the self-respect that a student possesses. This again relates back to the earlier section in this chapter that dealt with low motivated students who do not have high regard for themselves. Thus, if this is the case, the teacher must plan activities to help build self-esteem in these students in order to develop self-confidence and self-respect.

5. *Need for self-actualization.*—It is at this point in the application of Maslow's theory that true motivation and movement toward some goal can be realized by both the teacher and the student. After the first four needs have been met, the student can fully devote energies and efforts to accomplishing the objectives for that particular class. Furthermore, reaching this level encourages students to explore their minds and concentrate on acquiring new knowledge and learning skills for future use.

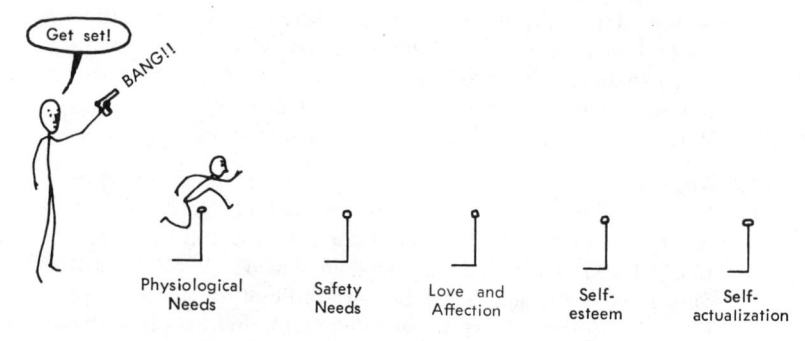

THE TEACHER MUST HELP EACH STUDENT
SATISFY THESE BASIC NEEDS

## Extrinsic Motivation

Extrinsic motivation focuses upon influences which occur from outside the student. This happens when some type of stimulus outside an individual's body tends to stimulate that individual to react. There are many different examples of extrinsic motivation that will be dealt with in the following section.

Certainly one of the most influential stimuli for extrinsic motivation is the teacher. The teacher is the one who controls the teaching / learning environment. Thus it is the teacher who can set the stage for creating a desirable atmosphere in which motivation can occur. In fact, it is the teacher's responsibility to assure that motivation does occur in each class to help move students toward a predetermined goal. Students will not come to class automatically motivated for the topic to be discussed. Thus the teacher must channel the student's energies and attention to the topic at hand and away from the upcoming football game, pep rally, or persons walking by the window or the door. With this in mind, there are certain tips which have been identified as practices which teachers can use to help provide extrinsic motivation for students.

1. *Much of the interest that an individual exhibits comes from natural impulses within that individual.*—While these natural impulses will tend to favor the intrinsic motivation of students, the teacher has many opportunities to provide extrinsic motivation to take advantage of the natural impulses. Examples of natural impulses are as follows:

    a. *Love of nature.*—Examples of pine cones, leaves, living plants, live animals, or other natural objects can help motivate students. It is not difficult to identify motivational objects which relate both to nature and to instructional topics at hand.

    b. *Curiosity.*—Curiosity implies that students' interests are aroused to the point that they are wondering why something did or did not happen, or they become curious when someone else knows the answer to something and they do not. One suggestion to help keep curiosity high is for the teacher to keep audiovisual aids, models, or other supplementary aids hidden from view until they are to be used.

    c. *Creativeness.*—All students have certain degrees of creative ability, and, if given the chance, they like to share this creativeness with others. The teacher should develop learning activities in which students are allowed to think creatively.

d. *Gregariousness.*—This implies that students, even though they might occasionally like to work by themselves, do prefer to be with others. This is why, when behavioral problems arise, students do not like to be removed from the classroom or laboratory. The point here is that students tend to respond better when they are with other students than when they are by themselves.

e. *Desire for approval.*—All students thrive on positive reinforcement. Whenever a teacher gives positive reinforcement, it tends to encourage students to repeat the action or strive for even further approval.

f. *Altruism.*—Even though some teachers may at first question this, most individuals generally have a concern for the welfare of others. A teacher can capitalize on this concern for others by using situations in which students are asked to help others or at least are given an opportunity to do so.

g. *Self-advancement.*—Students like to advance from one level of understanding to the next because they know they are moving toward some greater future goal. Students are more apt to work harder toward a distant goal when they know they are making advancement.

h. *Competition.*—Individuals like to compete with one another, and any time a teacher can develop a competitive session with the students, it makes it much easier to accomplish motivation. During this competition, learning can occur and points can be brought out by the teacher for the students to learn.

i. *Ownership.*—Perhaps this is one of the greatest assets of vocational teaching. Teachers can provide ownership to the students. The laboratory provides an excellent opportunity for the students to create new projects—to build, to repair, or to grow something. When the projects are completed, they belong to the students. The students can take them home and share with their parents the evidence that they did indeed learn something in a particular class.

## MOTIVATING STUDENTS

2. *Interest is contagious.*—If a teacher is able to get one or more students interested in something, their interest will help capture the interest of others. Basically, students feel they are missing something when other students are participating and they are left out. Thus, it is important that a teacher capture the interest of some of the students at the beginning of the class period in order for that interest to lead to the motivation of others.

3. *The novel or unexpected will arouse interest.*—The novel or unexpected is similar to curiosity which was mentioned earlier. Students basically like educational games or activities that are different. The creative teacher will add variety to the classroom activity by approaching class each day in a different way.

CURIOSITY WILL GO A LONG WAY
IN MOTIVATING A CLASS

4. *A subject is interesting if it affects students and others about them.*—Students always find themselves more interested or motivated in topics which affect them personally. This is especially valuable in the problem-solving approach in which the teacher provides opportunity for students to apply information that has been learned in the class or laboratory. This type of learning has much more meaning to students, and thus, when they leave school, they feel that they have learned something worthwhile when they can go home and apply what has been learned. This type of success and experience will provide students with a positive attitude when they return to school the next day and will help to stimulate motivation in future classes.

5. *Appropriate illustrations help.*—Students like to learn of previous experiences the teacher has encountered; therefore, it behooves the teacher to identify experiences that relate to the topic under discussion. This not only provides an opportunity for humor to be added to the class but also provides a source of excellent technical information which students may find useful.

6. *Satisfactory physical conditions are needed.*—This relates to Maslow's theory and specifically to the physiological needs. Learning will be hampered if the students must sit in a classroom or work in a lab where the temperature is uncomfortable. If there is a concern for safety, it is going to be very difficult for the teacher to teach or to motivate the students toward a particular topic. It is important that the teacher always check the surrounding conditions in order to minimize extremes which tend to hinder motivation of the students.

7. *Teaching should be based on thinking, not on memorizing.*—Memorization has become too common in our schools. Students are "turned off" when they are forced to memorize facts and figures which have very little use in securing future jobs. The teacher must avoid types of teaching situations in which useless memorization is required of the students. Students like to be challenged. Many different techniques can be used to help involve students in the teaching / learning process.

MOTIVATING STUDENTS 79

8. *Appropriate visual aids should be used.*—Students learn through their five senses—sight, sound, touch, taste, and smell. The use of any one or more of these different senses provides a great opportunity for the teacher to motivate the students' interest toward the topic under discussion. The use of only one sense will create a certain degree of motivation, the use of two a slightly higher degree, but the more senses involved in a situation, the more motivated the students will become.

INVOLVING THE FIVE SENSES IS IMPORTANT

9. *Teacher attitude.*—It is not uncommon in the literature to find that teacher attitude is related to the degree of motivation that students possess. The attitude of the teacher tends to indicate indirectly to the students the teacher's enthusiasm for the topic. If a teacher tends to be enthusiastic and exhibits this enthusiasm, the students will become more enthusiastic toward the topic under discussion.

In summary, in this specific section on motivation, the suggestions made are all extrinsic types of opportunities that a teacher can employ to help motivate the intrinsic nature of students. Even though a teacher might argue that the responsibility for motivation of any student rests within that student, the teacher has many powerful resources available which can be used to help motivate students. A responsible teacher uses these extrinsic factors in helping to motivate students toward individual and class goals.

## USING QUESTIONS IN MOTIVATING STUDENTS

It would not be appropriate to discuss motivation without devoting some attention to the role of the teacher in using questioning during motivational efforts. Regardless of which teaching technique is being used, it will involve questioning. In fact, it would be difficult for a teacher to begin a motivational section of a class without asking some questions. Motivation should occur not only at the beginning of a class, but also it should be present throughout the entire class period. It is the concern of the teacher to promote this motivation.

It is important to analyze this particular teaching technique and discover what is involved in constructing good questions.

Which type of question a teacher chooses depends upon the reason the question is being asked. Furthermore, critical thinking is an essential part of teaching, and questioning is one of the major tools to encourage and provide for the cognitive development of students. If students are expected merely to recall information for which only one answer is correct, then questions will need to be tailored as such. If students are expected to analyze a situation, then another type of question is needed. Following is a discussion of these four basic types of questions: recall, comprehension, analysis, and evaluation.

**MOTIVATING STUDENTS** 81

QUESTIONING IS AN IMPORTANT PART
OF THE TEACHING PROCESS

## Recall Questions

Recall questions are to be answered with specific facts, principles, or generalizations, for example, dates, events, persons, places, principles. In constructing questions for which this information is desired, the teacher should begin the question with *who, what, when,* or *where*. Teachers should not use this type of question to an excess because it represents a low level of thinking. Recall questioning can be used to guide students to a high level of intellectual thinking. Examples of this type of question are:

Who is the National FFA president?
What is the weight of a bushel of shelled corn?
When should Easter lilies be planted?
Where can parts for a small engine be purchased?

## Comprehension Questions

Comprehension questions are to be answered by indicating an understanding of something. This question demands manipulation of data through interpretation, summarization, example, and definition. Interpretations require students to show likenesses, differences, or comparisons; summarizations require students to restate ideas in their own words. Examples call for an illustration of the ideas being discussed; definitions provide the opportunity for the students to develop their own definitions. Key words to use in stating comprehension questions are *how*, *why*, *compare*, or *give*. Examples of this type of question are:

How would you define public speaking?

Why are farms becoming larger in size?

Compare the operating costs of gasoline and diesel engines.

Give an example which indicates that soil conservation must be practiced.

## Analysis Questions

Analysis questions require students to separate data for the purpose of discovering hidden meaning, relationships, or basic structure. Furthermore, analysis questions should be used to seek out underlying relationships and organizational patterns. Key words to use must suggest analysis and these words are *relationships, assumptions, motives, implications,* or *identification of issues*. Analysis questions must be considered in light of established criteria. Examples of analysis questions are:

How are plant diseases related to environmental conditions in the greenhouse?

What are the implications of exceeding the recommended planting rate?

What is the motive behind the banning of pesticides?

## Evaluation Questions

Evaluation questions provide the opportunity for students to make judgments, opinions, personal reactions, and criticisms based upon their own criteria. When using evaluation questions, there are no right or wrong answers, but the students are expected to defend their answers, and their answers serve as the basis for judging or evaluating the students' responses. Evaluation questions involve the intellectual and emotional aspect of the students, thus at times, responses may be highly biased. Key words to use are *should, could, would,* or *in your opinion.* Examples of questions are:

>Should the youth club go on a two-week summer trip?
>
>In your opinion, should Florist Jones buy Hybrid A or Hybrid B?
>
>What is your personal reaction to the stocking of wild turkeys in this area?

In summary, when preparing to ask a question, the teacher must give some thought as to why it is being asked and then decide how to state it. Effective questioning as a part of the motivational plans of the teacher takes practice, and the teacher may find it necessary to write out questions until the ability to ask quality questions spontaneously is developed.

# CHAPTER 4

# ▶ Selecting Teaching ◀
# Techniques

ONE OF THE FIRST QUESTIONS TO ASK when getting ready to select a teaching technique is "Just what happens when a teacher teaches a class?" For example, think back to a recent class, and describe exactly what went on during that class period. Search out answers to the following questions:

- What did the teacher do?
- What did the students do?
- What types of questions were asked?
- What visual aids did the teacher use?
- What student activity occurred?
- Was the class formal or informal?
- How did the teacher's personality enter into the teaching?
- What non-verbal cues were observed?

As a person begins to analyze what goes on in a teaching / learning environment, it is quite evident that teaching is a complex activity in which many factors influence the success or failure of the class in achieving the established objectives. The problem which this chapter will address deals with the selection of teaching techniques. This is one decision that a teacher cannot take lightly, and the manner in which the teacher selects a teaching technique will ultimately influence whether the appropriate technique was selected for a particular class.

There are certain terms which should be clarified before any discussion is held on selecting teaching techniques. These terms are *teaching strategy, teaching techniques,* and *instructional aids.* Each of these terms will be discussed in greater detail in the following sections.

## TEACHING STRATEGY

The military service has long talked about strategies in planning battles or other military engagements. We also find the word "strategy" being used during football games and hear TV announcers talk about the strategies that golfers contemplate as they approach the green. Educators must also be thinking about teaching strategies as they plan instructional classes. Strategy, as it relates to teaching, refers to the manner in which the class is conducted and to the content presented so that the ultimate goal or objective of the class is achieved. To elaborate further, the teaching strategy is the sum total of everything a teacher plans and carries out in order to reach a set goal or objective. Two teachers, having the same objectives, may teach the same subject to two different classes using two different approaches, and yet both may achieve their objectives. Another way of viewing teaching strategy is that it is really a mixture of the teacher's personality, motivation used, teaching techniques involved, instructional aids used to supplement the class, student involvement, and non-verbal cues used during the class. In summary, teaching strategy is everything

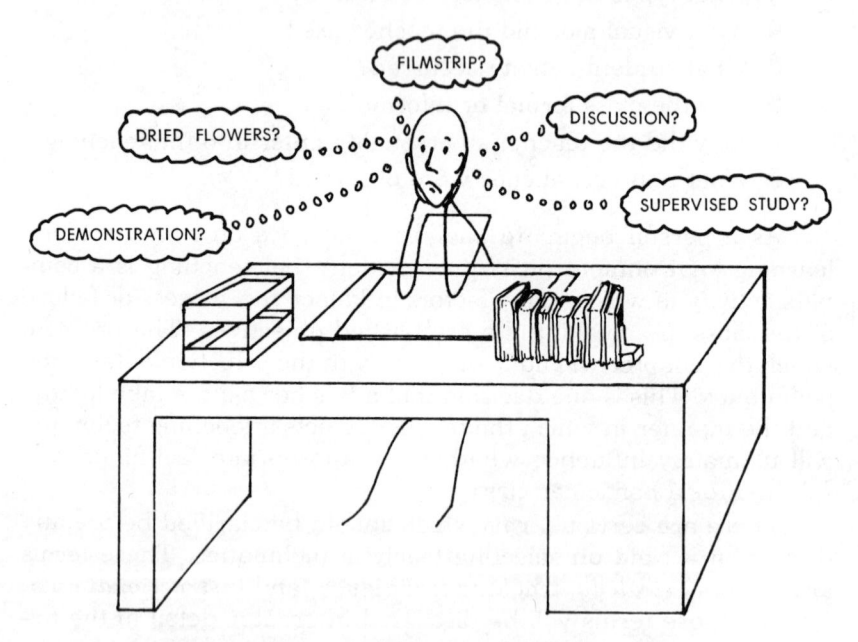

DECIDE HOW TO TEACH THE CLASS BEFORE IT BEGINS

that a teacher's creative mind can assemble and arrange in an orderly manner so that the class becomes an effective learning situation through which the ultimate goals of the class are realized.

Different teaching strategies generally defy exact categorization. For a beginning teacher, it is not important to try to distinguish one strategy from another. Each teacher should, within a specified class period, use all human and material resources available in order to help assure that the class is a success.

## TEACHING TECHNIQUES

Teaching techniques refer to specific approaches to teaching that teachers may use to help convey information to students so that the students are able to achieve the ultimate goal or teaching / learning objectives. Teaching techniques are, in most cases, something teachers cannot put their hands on. For example, teachers cannot put their hands on supervised study or team teaching. Teaching techniques are something that teachers must manipulate (not necessarily physically) with professional skill in order to do what they want them to do. In teaching, this implies that teaching techniques are something teachers use to aid in conveying certain information to students.

## INSTRUCTIONAL AIDS

Instructional aids are objects which supplement and / or complement the learning environment so that the ultimate goal can be achieved more easily. Instructional aids are aids that teachers can manipulate manually during the teaching / learning process to help clarify instructional content, convey information to the students, or review previously covered content. However, it is not always easy to distinguish clearly between teaching techniques and instructional aids. For example, is videotape a teaching technique or an instructional aid?

Actually, it depends on how the teacher uses this particular hardware. For example, if the teacher uses the videotape machine to tape students who are preparing for an FFA judging contest, the use of videotape is a teaching technique. However, if the teacher is teaching exactly the same material to two different classes and tapes the first class's presentation for use in the later class, then the videotape process becomes more of an instructional aid, similar to a carousel projector with a set of slides. In summary, the difference between a teaching technique and an instructional aid depends on how the teacher uses these resources in a specific teaching situation.

## TEACHING TECHNIQUES AND PROBLEM SOLVING

Perhaps a question to be asked now is, "What is the difference between problem solving and teaching techniques?" This is an area in which clear definitions have been neither agreed upon nor accepted by the teaching profession. Problem solving can be thought of as a method of teaching in which the teacher guides the class through a series of questions which, when answered by the students, will provide the basis for enhancing the teaching / learning process. Teaching techniques are specific tools which can be used within the problem-solving approach. As it was demonstrated elsewhere in this book, problem solving is a method used by the teacher to identify problems that need to be addressed within a certain instructional unit. After these problems have been identified, the teacher may use different teaching techniques in order to solve these problems and to provide each student with the basis for making final decisions as to what is the best solution to the problems. Problem solving is different than teaching strategy. While a teaching strategy encompasses how a teacher blends teaching techniques, aids, and learning activities over the entire class period or several class periods, problem solving tends to permeate the teaching strategy with questions posed by the teacher which when answered will result in student learning.

## FUNCTIONS OF TEACHING TECHNIQUES AND INSTRUCTIONAL AIDS

To make effective use of the teaching techniques and the instructional aids available to them, it is essential that teachers have a sound concept of the different functions that teaching techniques and instructional aids can assume in the teaching / learning environment. Discussions of these unique functions follow.

## Conveying Technical Information

One of the most common functions of teaching techniques is that of conveying technical information to students. Without a doubt, this is one of the basic functions of both teaching techniques and instructional aids. This function can be achieved through a variety of teaching techniques, and different teachers may select different techniques for conveying similar technical information.

## Promoting Student Interest

Certain teaching techniques lend themselves to promoting or developing student interest in a particular topic. For example, field trips are generally used during the last part or at the end of an instructional topic in order to bring the lesson to a close. As another strategy, field trips could be used at the beginning of a topic in order to promote or stimulate student interest in the topic. Experiments, games, or other teaching techniques could be used to help stimulate or motivate student interest. One of the common forms used by teachers in promoting interest centers around the discussion technique. However, an effective teacher will vary the approach from time to time and use different teaching techniques to help to stimulate and maintain student interest.

## Helping Students Retain Information

Research has shown that in situations in which a teacher continually used previously learned information in different ways, the information learned by the students was retained for a longer period of time. Thus, one function of teaching techniques is to provide the teacher with the opportunity to use different approaches in order to enhance student learning. Also, topics can be reviewed or summarized by using different techniques. As an example, perhaps supervised study was used to cover a particular topic, and, three days later, the teacher felt a review of the material was necessary. A game could be utilized in which the students would be required to recall previously learned information. Rather than go back to a typical question / answer period for a review, the game as a teaching technique could serve to help students retain what they had learned earlier. Another popular function of teaching techniques would be to have the students apply the knowledge they had gained in the classroom to laboratory or greenhouse projects or experiments. The basic premise of this particular function is that it requires students to recall or apply what they had already learned and that through this continual use of information, they will be better able to retain what they had learned earlier.

## Changing Teaching Strategy

Without a doubt, education would be very dull if teachers continually chose to approach their classes in the same manner day after day. One of the best uses of teaching techniques is to provide teachers with the opportunity to vary their strategies in approaching instructional topics. Teachers can vary their approach through different teaching techniques and thus prevent boredom from setting in.

## Making Teaching More Enjoyable

While the preceding functions focused upon the student, different teaching techniques can also help to motivate teachers and prevent their becoming stagnant. Teachers have a habit of becoming set in one particular strategy or technique, and, period after period, they might begin to rely upon the same teaching technique and instructional aid. As an example, discussion is a technique commonly used and many times discussion could be avoided and another teaching technique used in its place. A continual change in strategies and teaching techniques will help to motivate teachers and keep them more enthusiastic about the topics under discussion. Another point to make at this time is that within any one day, a teacher may be assigned to teach six different classes. Certain techniques require different amounts of teacher input, and thus some teaching techniques allow the teacher to operate on a more informal basis. In this situation, the pressure on the teacher from being in front of the class for a full 55-minute period can be relieved. If a teacher uses supervised study in a 55-minute period and the time set aside for student reading is 15 to 20 minutes, this is 15 to 20 minutes that the teacher can circulate around the classroom at leisure and help those students who need special assistance. The point to be made here is that if a teacher is faced with teaching several straight periods, teaching will become easier and more enjoyable if teaching techniques are varied throughout the school day.

TEACHER ENTHUSIASM IS CONTAGIOUS

## Involving the Five Senses of Hearing, Seeing, Tasting, Touching, and Smelling

A critical review of different teaching techniques will immediately point out that a certain technique will use one sense while another technique will use another sense. This review of the teaching techniques and instructional aids will also point out that certain techniques and aids require the use of more than one sense. It must be remembered that if the teacher continually uses one particular teaching technique, then the student is forced to use only certain senses for learning purposes. For example, in a classroom discussion, the student just needs to listen. Thus, if a teacher continually uses discussion day after day, the student will just listen without using any other senses. However, if an experiment is used, the student may, in addition to listening, also be required to see, taste, touch, or smell something as the experiment is conducted. Research has shown that the five senses are very important if successful learning is to occur, and, from time to time, the teacher needs to involve teaching techniques and aids that will require the use of different senses by the student during the learning process.

### Guiding the Learning Process

Certain teaching techniques tend to guide students in the learning process. Thus, it relieves the teacher from the pressure of being in front of the class continually directing student thinking. For example, a supervised study situation can be used to help guide the student learning process by which students can move individually from one point to the next higher order of thinking or level of understanding. Supervised study does not require the teacher to be in front of the class every minute, thus being the major focus of the learning environment. Programmed instruction is another example of a teaching technique that helps to move the student from one point to the next point with little or no direction from the teacher.

### Supplementing Other Teaching Aids and Techniques

As stated in the earlier description of teaching strategy, very few teaching techniques and instructional aids are used alone. Many times they are used in conjunction with other teaching techniques and aids. The use of supplemental teaching techniques and aids is just another way a teacher can help to form a teaching strategy that is intended to promote the student's success in reaching established goals and objectives.

### Assessing What Has Been Learned

While teaching techniques can be used to convey information, they can be used also to help the teacher in assessing the learning achieved in previous classes. Games, role playing, field trips, contests, and questioning are but a few of the teaching techniques which can be used to measure student achievement.

## DECIDING WHICH TEACHING TECHNIQUES TO USE

There are many different techniques available, and the teacher should consider each one as a possibility when planning an instructional lesson. One of the first steps in selecting teaching techniques for a particular instructional unit is to identify them.

## SELECTING TEACHING TECHNIQUES

The teaching techniques that meet the definition discussed earlier in this chapter are:

| | |
|---|---|
| Panel discussions | Brainstorming |
| Experiments | Educational TV |
| Programmed instruction | Testing |
| Team teaching | Contests |
| Micro teaching | Games |
| Videotapes | Discussions |
| Individualized instruction | Supervised study |
| Seminars | Student reports |
| Debates | Role playing |
| Questioning | Demonstrations |
| Field trips | Resource persons |

These particular teaching techniques will not be discussed in depth. There are many books on the market that deal with these techniques and the unique characteristics that each one possesses. The following discussion will focus upon how a teacher can select the particular teaching techniques suitable for a specific situation.

*Let's see what I pull out of the hat today.*

PULLING TEACHING TECHNIQUES OUT OF THE HAT IS FOR DR. SEUSS

## Formula for Selecting Teaching Techniques and Instructional Aids

While the selection of teaching techniques and aids is not an exact science, a suggested procedure to assist the teacher in developing a thought process for the selection is offered here.

$$\text{Lesson topic} + \text{Behavioral objective} + \text{Factors} = \text{Teaching techniques} + \text{Instructional aids}$$

Following, each part of this formula is discussed.

*Lesson Topic.*—The information for this particular part of the formula will come from the teaching calendar or course of study which the teacher already has developed. For example, this may be for sketching a project plan, gapping a spark plug, or fertilizing corn. The identification of the lesson topic begins to focus upon both the instructional unit to be taught and the expected behavior of the student.

*Behavioral Objective(s).*—The next part of the formula provides the teacher with the opportunity to identify specifically the behavioral objective(s) for the class. Furthermore, these behavioral objective(s) should pinpoint the final behavior expected of the students after they have completed the lesson. In a review of many resources dealing with behavioral objectives, the terms that a teacher selects to put into this part of the formula must be very specific, and words must be avoided which are prone to a wide variation of interpretation. For example, the teacher would want to avoid the terms, *understand, know,* and *appreciate.* For each of the learning domains (cognitive, affective, and psychomotor), samples of the verbs which can be used have been listed in Figures 4-1, 4-2, and 4-3.

*Dealing with Factors.*—No teaching environment is an ideal situation. When determining which teaching techniques might be most appropriate, the teacher will have to consider certain factors that will influence the type of teaching techniques and aids that eventually can be selected. Some of the common factors that will need to be considered are the following:

- The abilities and interests of the group
- Size of the class
- Teacher competencies

SELECTING TEACHING TECHNIQUES

## FIGURE 4-1
### Selected Action Verbs Appropriate to the Domains of the Taxonomy:

### COGNITIVE

| Knowledge | Comprehensive | Application |
|---|---|---|
| acquire | associate | apply |
| count | classify | calculate |
| define | compare | change |
| draw | compute | classify |
| identify | contrast | complete |
| indicate | convert | demonstrate |
| label | describe | discover |
| list | differentiate | employ |
| match | discuss | examine |
| name | distinguish | illustrate |
| outline | estimate | manipulate |
| point | explain | operate |
| quote | extrapolate | practice |
| read | interpret | prepare |
| recall | interpolate | produce |
| recite | predict | relate |
| recognize | rewrite | solve |
| record | translate | use |
| repeat | | utilize |
| state | | |
| tabulate | | |
| trace | | |
| write | | |

| Analysis | Synthesis | Evaluation |
|---|---|---|
| analyze | arrange | appraise |
| construct | categorize | assess |
| detect | combine | compare |
| diagram | construct | critique |
| differentiate | create | determine |
| explain | design | evaluate |
| infer | develop | grade |
| outline | explain | judge |
| separate | formulate | justify |
| subdivide | generate | measure |
| summarize | generalize | rank |
| | integrate | rate |
| | organize | recommend |
| | plan | select |
| | prepare | support |
| | prescribe | test |
| | produce | |
| | propose | |
| | rearrange | |
| | reconstruct | |
| | specify | |
| | summarize | |

## FIGURE 4-2
### Selected Action Verbs Appropriate to the Domains of the Taxonomy:

#### AFFECTIVE

| Receiving | Responding | Valuing |
|---|---|---|
| accept | answer | argue |
| accumulate | approve | assist |
| ask | commend | debate |
| choose | comply | deny |
| combine | conform | help |
| control | discuss | increase measured proficiency in |
| differentiate | follow | |
| follow | help | increase numbers in |
| listen (for) | play | join |
| reply | practice | protest |
| select | read | read |
| separate | volunteer | relinquish |
| set apart | | select |
| share | | specify |
| | | support |

| Organization | Characterization |
|---|---|
| abstract | act |
| adhere | avoid |
| alter | change |
| arrange | complete |
| balance | display |
| combine | manage |
| compare | perform |
| define | quality |
| discuss | rated high by peers |
| formulate | require |
| integrate | resist |
| organize | resolve |
| prepare | revise |
| theorize | serve |
| | solve |
| | superiors, or subordinate |
| | verify |

## FIGURE 4-3

Selected Action Verbs Appropriate to the
Domains of the Taxonomy:

### PSYCHOMOTOR

| | |
|---|---|
| apply | make |
| assemble | make-up |
| build | manipulate |
| calibrate | manufacture |
| change | measure |
| clear | mix |
| compose | operate |
| connect | organize |
| construct | perform |
| correct | plan |
| cut | position |
| demonstrate | put together |
| design | remove |
| desire to respond | restore |
| discover | see |
| dismantle | select |
| draw | sense |
| fabricate | service |
| fasten | sharpen |
| feel | simulate |
| follow | smell |
| form | touch |
| gap | trace |
| hear | troubleshoot |
| imitate | try |
| install | recognize |
| lay out | use |
| locate | visualize |
| maintain | |

▶ Facilities available
▶ Time available
▶ Behavioral objective of the class
▶ Nature of the subject matter
▶ Resource material available
▶ Nature of the teaching technique

Following, a few of these factors are explained in greater detail.

*Size of the Class.*—The number of students in a class influences the selection of certain types of teaching techniques and aids. For example, in order to have a good educational game, a certain number of students may be required to operate the game effectively. Thus, if the size of the class is not appropriate for the game in mind, then that game as a teaching technique would have to be eliminated. Also, in order to present an effective demonstration for a class of 25 students, the teacher might have to divide the class into small groups to assure that all 25 students would be able to see and hear the demonstration in order to derive full benefit from it.

*Teacher Competency.*—A teacher may be very competent in the use of several techniques, but the habit of always using several of the same techniques and excluding all others can affect the teacher's ability to select the most effective teaching technique for a certain lesson. For example, a teacher may realize that an experiment would be the best teaching technique to use in order to get the content across to the students. However, if the experiment technique is one which had never been used, the teacher would be reluctant to use that technique because of this inexperience. Indirectly, this lack of competency would tend to discourage the use of experiments as a teaching technique for any lesson under consideration by that teacher.

*Time Available.*—Time is one factor that is a constant concern of the teacher. A discussion may be the best way to approach a certain instructional unit, but discussion can be very time consuming. Thus, if there is a shortage of time, a more formal, teacher-controlled teaching technique may be better to help move the students along the topic at a rapid pace.

*Availability of Resource Materials.*—The selection of certain types of teaching techniques and aids will be influenced by the availability of resource materials. Supervised study essentially requires some type of reading material on the part of the student. Thus, if ample materials are not available for each student, then this particular teaching technique would be of limited use to the teacher. This would also be true for experiments for which certain resource materials are needed by the student in order to conduct an effective experiment.

*Nature of the Teaching Technique.*—Field trips require

## SELECTING TEACHING TECHNIQUES

movement out of the classroom or laboratory into the school surroundings or even into the community. There are certain types of instructional topics that lend themselves very well to field trips while others would be best not approached through the use of this technique.

Another factor that would influence the type of teaching technique used is the degree of control that the teacher wants to exert over the students. Some teaching techniques require the teacher's approach to be direct or formal. Questions, discussions, and lectures are all examples of situations in which the teacher is usually in command of the class. On the other extreme, there are some teaching techniques, such as programmed instruction, role playing, and using resource persons, by which the teacher adopts a passive or informal role. Points to be made here are that if an instructional topic under consideration is one in which the teacher wants to develop the student's ability to assume leadership qualities, an informal type of teaching technique should be used. If the particular topic is one which, for safety or other reasons, the teacher needs to be in greater control of the class, a formal approach would be more appropriate.

If a teacher were to follow through the thought process inherent in the procedure as presented in this chapter, certain teaching techniques should begin to surface and be used, or at least considered, in teaching a particular topic. Furthermore, instructional aids or materials essential for an effective class can be identified also through the suggested procedure. This can be illustrated by the following example in which "Gapping a Spark Plug" is the lesson topic.

$$\text{Gapping a spark plug} + \text{Gapping} + \text{Factors}$$
$$= \text{Demonstration} + \begin{cases} \text{Spark plug} \\ \text{Gapping tool} \\ \text{Owner's Manual} \end{cases}$$

As can be seen from this example, the behavioral objective becomes one of gapping, a psychomotor type of activity. The factors part of the formula would need to be considered on a more individualized basis for those situations currently existing within a local school. In this case, the factors to consider would be class size, teacher competency, time available, resources available, and objectives. Following the formula, the teaching technique that would help the student best learn to gap a spark plug would be a demonstration in which a teacher would actually go through the steps in gapping a spark plug followed up with actual student application. It is quite obvious that in order to teach this topic, the instructional aids would be a spark plug, gapping tool, and an owner's manual that would describe the appropriate gap that a spark plug should have for a particular power unit.

Another consideration in the selection of techniques concerns the particular strengths or weaknesses that each technique possesses. For example, some techniques involve a large number of students while other teaching techniques limit the number of students. Here again, this is pointed out through several illustrations. Role playing is a teaching technique used quite frequently in the agribusiness instructional area. If a unit is under consideration that deals with interviewing, two students could very well assume the roles of the interviewer and interviewee. However, this involves only two students in the class; therefore, the teacher must either provide similar roles for other students in other parts of the classroom or provide other educational activities to occupy the students while two of the students role play. One suitable activity is for the class to observe and evaluate the role playing. Educational games often are used, and these games should involve all of the students in the class, not just a select few. Another factor to consider in the use of educational games is that any game that eliminates students from participating is not a desirable technique to use. If the first several students to lose have to sit out the rest of the game, they will soon discover that spending the rest of the period just watching other students still actively involved in the classroom activities is not very interesting. The point to be made here is that each teaching technique has particular strengths or weaknesses, and all teachers must be aware of these as they select appropriate teaching techniques for the lesson topic under consideration. It is only through this logical thought process that teachers will be able to avoid many problems that some teachers experience in trying to teach a particular class while using an inappropriate technique. Specific evaluation of selected teaching techniques is discussed further in Chapter 12.

A second example, in which "Purchasing Building Supplies" is the lesson topic, provides a different situation for the teacher. Applying the formula, it would look like this:

## SELECTING TEACHING TECHNIQUES

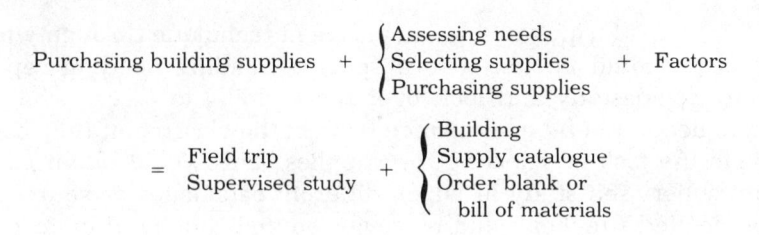

Purchasing building supplies + $\begin{cases} \text{Assessing needs} \\ \text{Selecting supplies} \\ \text{Purchasing supplies} \end{cases}$ + Factors

= $\begin{matrix} \text{Field trip} \\ \text{Supervised study} \end{matrix}$ + $\begin{cases} \text{Building} \\ \text{Supply catalogue} \\ \text{Order blank or} \\ \quad \text{bill of materials} \end{cases}$

In this case, several teaching techniques should be used by the teacher. The basis for this reasoning is founded in the number of tasks that a student would need to complete in order to purchase supplies for the maintenance of a building. These tasks are:

▶ Assessing building supplies
▶ Selecting supplies
▶ Purchasing or ordering supplies

TEACHING TECHNIQUES SHOULD ASSIST STUDENTS
IN SOLVING PROBLEMS

The field trip would be an excellent technique through which students could be taken to a specific building needing repair. They could study and look over the building to assess what repairs needed to be made. Once back in the classroom, they could begin the task of selecting the supplies. This would be an excellent supervised study in which different catalogues or sources of the needed supplies could be reviewed and compared in order to make the most economical purchases. Achieving the next task (ordering or purchasing supplies) could be an application by which students would actually prepare a bill of materials and place the order with a local business. This might be taught through the use of several different techniques. The teacher must keep in mind that the overall lesson topic and behavioral objective will not by themselves give complete clues to the types of teaching techniques or instructional aids needed. A thorough analysis of the lesson topic in which consideration is given to specific tasks or competencies that students are to master will point out where more than a single teaching technique will be needed to teach the material successfully.

# CHAPTER 5

# Reviewing Some Illustrative Plans

IT IS OBVIOUSLY IMPOSSIBLE for anyone except the teacher to develop suitable plans for a particular class. The following plans are included only in the interest of developing an adequate understanding of this presentation of a problem-solving approach in teaching.

The plans as presented here are incomplete. Only enough is being presented to illustrate each phase of the plan. Discussion has been limited purposely in order to obtain a clear picture of the plan as it should be developed.

### PLAN NUMBER ONE (A PROBLEM AREA PLAN)
**Part I. Pre-planning Analysis**
    A. Enterprise or activity: Soils and fertilizers
    B. Problem area: Making and using soil tests
    C. Analysis of the teaching situation:
        1. The class has had instruction in taking soil samples.
        2. About 10 percent of the farmers are making use of a local soil testing bureau.
        3. Soils are deficient in all major elements—nitrogen, phosphorus, and potash; lime also is needed.
        4. Two of the home farms represented in the class have limited soil testing programs in operation.
        5. Soil testing is of interest to the entire class as an area of knowledge important in several agricultural occupations—teaching agriculture, farming, farm / range management, fertilizer sales, etc.
    D. Teaching / learning objectives: To develop the ability of the students to make soil tests and to use the results of soil tests in connection with their homes or supervised agricultural experience programs. Specifically, to develop these abilities:

1. To determine what soil tests should be made.
2. To determine what use should be made of the local soil testing bureau.
3. To make soil tests for pH, phosphorus, potash, and calcium.
4. To interpret soil test results.
5. To make a map showing the soil test results for the fields tested on farms.
6. To determine the fertilizer needs of the fields tested as indicated by the test results and previous history of crops and fertilization on the fields.
7. To identify the occupations in which a knowledge of soil testing would be used.

**Part II. The Teaching Plan**
   A. Enterprise or activity: Soils and fertilizers
   B. Problem area: Making and using soil tests
   C. Interest approach (discussion questions):
      1. Who uses soil tests?
      2. How are soil tests used on a farm? In other businesses?
      3. What experiences have you had in making soil tests?
      4. How could you tell whether using soil tests has been a profitable practice?
   D. Anticipated group objectives:
      (Lead question: Why is it important that we be able to do a good job of making and using soil tests?)
      1. To keep from running down our soil.
      2. To get yields of 150 bushels of corn per acre; 40 bushels of soybeans per acre; and 25 tons of silage per acre.
      3. To make the best use of the money we have available for buying fertilizer.
   E. Anticipated problems and concerns of the students:
      (Lead question: What things do we need to know and be able to do in order to make and use soil tests well and accomplish our objectives?[1]
      1. Who makes soil tests?
      2. What soil tests should be made?
      3. How is each test made?
      4. How are the test results reported?
      5. What do the tests mean in terms of the need for fertilizer?

---

[1] Further questioning of a specific nature will be needed to draw out some of the problems. These questions are developed by the teacher in response to the situation as the discussion moves along.

6. How are the test results shown on maps of the fields tested?
7. How often should the fields be tested?
8. How much does it cost to test soil?
9. What equipment and supplies are needed for testing soil?
F. Steps in solving the problems:
1. Have the students select a problem from the list. (Similar and overlapping problems may be taken up together.)
2. Lead the students in a discussion of the problem to find out what they do and do not know.
3. Conduct supervised study on those things the students do not know.
4. Lead the students in the final discussion and drawing of conclusions. (It may be necessary to have more supervised study if the discussion reveals more questions that the students cannot answer.)
5. Have the students select another problem from the list and repeat the above steps.
G. Evaluation and application:
1. Develop a list of approved practices with the class.
2. Have the students write plans for using their knowledge of soil testing and soil tests in their individual agricultural experience programs.
3. Give a written examination. (The examination should be constructed as a part of the teaching plan.)
4. Make plans for needed follow-up instruction in connection with student agricultural experience programs.

### Part III. Teaching Resources

A. References
B. Teaching aids
C. Teaching techniques

## PLAN NUMBER TWO (A PROBLEM AREA PLAN)

### Part I. Pre-planning Analysis

A. Enterprise or activity: Corn
B. Problem area: Planting corn
C. Analysis of the teaching situation:
1. Corn is grown for grain on nearly all of the farms of the community.
2. Several of the class members are working in nonfarm agricultural businesses which sell seed corn or sell and service planting equipment.

3. The class has had instruction in selecting and buying seed corn.
4. The class has had instruction in making and using soil tests.
5. The farmers in the community appear to be doing a good job of planting corn.
D. Teaching / learning objectives: To develop these abilities:
1. To check equipment needed for planting and to make adjustments as they are needed.
2. To determine the proper rate of planting.
3. To set the corn planter to sow at the proper rate.
4. To prepare the seedbed for planting.
5. To apply the proper amount of fertilizer.
6. To prepare the seed corn for planting.
7. To determine the proper time to plant.
8. To apply the proper amounts of chemicals for controlling weeds and insects.

**Part II. The Teaching Plan**
A. Enterprise or activity: Corn
B. Problem Area: Planting corn
C. Interest approach (discussion questions):
1. How many of you have ever planted corn or helped to plant corn?
2. Just how did you go about it?
3. How did you decide how well the task of planting was done?
4. Who, besides the farmer, needs to know how to plant corn?
D. Anticipated group objectives:
(Lead questions: What were the goals we decided upon for corn yields when we studied about soil testing and selecting and buying seed corn? Are these still good goals toward which to work?)
1. To get at least 150 bushels of corn per acre.
2. To plant corn in straight rows and get an even stand.
3. To get 20,000 plants per acre.
E. Anticipated problems and concerns of the students:
(Lead question: You have mentioned some of the things we have to be able to do to plant corn. Now let's see if we can list all of the things we need to know and be able to do in planting corn in order to get the yields we decided upon as our goals.)
1. What equipment is needed and how is it adjusted?
2. How are the fields prepared for planting?

# REVIEWING SOME ILLUSTRATIVE PLANS

PLANNING WILL PREPARE THE TEACHER
FOR ANY SITUATION

3. What should be the rate of seeding per acre?
4. How is the corn planter adjusted to sow at the right rate?
5. How is the seed prepared for planting?
6. How is the fertilizer applied?
7. When should the corn be planted?
8. How are weed and insect chemicals applied at planting time?

(In some instances, the problems listed by the students may have been considered in connection with some other problem area. In cases like this, the question will be used for review and may not require special supervised study.)

F. Steps in solving the problems:
1. Have the students select a problem from the list.
2. Lead the students in a discussion of the problem to find out what they do and do not know.
3. Conduct supervised study on those things the students do not know.

4. Lead students in the final discussion and drawing of conclusions.
5. Have students select another problem from the list and repeat the above steps.
G. Evaluation and application:
   1. Develop a list of approved practices.
      a.
      b.
      c.
      d.
   2. Have the students write plans for supervised farming programs.
   3. Test (develop in advance).
   4. Follow-up instruction on supervisory visits.

**Part III. Teaching Resources**

A. References
B. Teaching aids
C. Teaching techniques

IT IS IMPORTANT TO ALWAYS BE PREPARED WITH A GOOD TEACHING PLAN

## PLAN NUMBER THREE (A UNIT PLAN)

**Part I. Pre-planning Analysis**

A. Enterprise: Swine
B. Unit: Controlling diseases and parasites of swine
C. Problem areas:
   1. Preventing and controlling diseases of swine
   2. Controlling parasites of swine

D. Analysis of the teaching situation for the unit:
   1. Hogs are grown on over half of the farms in the area.
   2. Feed and livestock supplies salespersons are often asked about swine diseases and parasites.
   3. The class has had instruction on selecting, buying, and feeding hogs and on providing housing and equipment for them.
   4. The class is in the second semester of the freshman year.
   5. There have been no major outbreaks of hog diseases recently. There has been trouble with erysipelas and TGE.
E. Teaching / learning objectives for the unit: To develop the ability:
   1. To recognize common diseases and parasites and their symptoms.
   2. To use proper practices to prevent diseases and parasites from invading the swine herd.
   3. To apply necessary measures for controlling diseases and parasites.
   4. To plan a disease and parasite prevention program for a farm on which swine are raised.

## Part II. The Teaching Plan

A. Enterprise or activity: Swine
B. Unit: Controlling diseases and parasites of swine
C. Problem areas:
   1. Preventing and controlling diseases of swine
   2. Controlling parasites of swine
D. Interest approach for the unit:
   1. Explain the major disease and parasite problems hog producers are facing in this area.
   2. What is a disease? Parasite?
   3. What other diseases and parasites were mentioned in the other units and problem areas on swine?
   4. What practices are being used to prevent and control swine diseases and parasites?
   5. How can one judge how good a job is being done in disease and parasite prevention and control?
E. Anticipated group objectives for the unit:
   (Lead question: Why is it important that we do a good job of preventing and controlling diseases and parasites in swine?)
   1. To reduce death loss to nearly zero.
   2. To wean pigs at four weeks of age.
   3. To get hogs to market weight of 200 pounds in five months.
   4. To have a swine herd of which we can be proud.

F. Study of problem areas:
  1. Problem area #1: Preventing and controlling diseases of swine
     a. Discussion of unit objectives which apply.
        (Lead question: Which of the unit objectives apply to this problem area? How?)
     b. Anticipated problems and concerns of the students:
        *Stage 1.*[2] (Lead question: What diseases is it important that we know about? Start with those named in the interest approach.)
        *List of diseases* (problems)
        (1) Hog cholera
        (2) Brucellosis
        (3) Erysipelas
        (4) TGE
        *Stage 2.* (Lead question: What do we need to know and be able to do concerning each disease?)
        *List of questions* (guide questions for study)
        (1) What are the symptoms of the disease?
        (2) What is the cause of the disease?
        (3) How does the disease spread and get into a herd?
        (4) How can the disease be prevented?
        (5) How can we control the disease?
     c. Steps in solving the problems:
        (1) Select a disease.
        (2) Discuss to determine what is and is not known.
        (3) Study to check on the "known" and to find answers to the unknown.
        (4) Discuss to arrive at conclusions.
        (5) Select another disease and repeat the process.
     d. Approved practices:
        (Instead of the usual list of practices, the preventive measures for the disease studied may be organized into a disease prevention program for a farm on which hogs are raised.)
  2. Problem area #2: Controlling parasites of swine
     a. Discussion of unit objectives which apply.
        (Lead questions: Which of the group objectives apply to this problem area? How? Are there any others we should recognize?)

---

[2]This kind of problem area is most effectively treated by using a two-stage approach. The first stage consists of identification of the diseases which will be treated as if they were problems; the second stage consists of what needs to be learned about each disease, or guide questions for study.

b. Anticipated problems and concerns of students:
   *Stage 1.* (Lead question: What parasites is it important that we know about, including those named in the interest approach?)
   *Parasites*:
   (1) Roundworms
   (2) Lice
   *Stage 2.* (Lead question: What do we need to know and be able to do concerning each parasite?)
   *List of questions* (guide questions for study)
   (1) What are the stages in the life cycle of the parasite?
   (2) How can its presence be determined in the herd?
   (3) How does the parasite get into the herd?
   (4) How can it be prevented from getting into the herd?
   (5) How can the parasite be eliminated from the herd?
c. Steps in solving the problems:
   (1) Select a parasite.
   (2) Discuss to determine what is and is not known.
   (3) Conduct supervised study.
   (4) Discuss to develop conclusions.
   (5) Select another parasite and repeat the process.
d. Approved practices: A parasite prevention and control program
G. Evaluation and application for the unit:
   1. Plans for supervised agricultural experience programs.
   2. Plans for on-the-job instruction.
   3. Test.

## Part III. Teaching Resources

A. Resources for problem area #1
   1. Assigning the study of specific diseases to small groups or individuals. Some teachers prefer this procedure if study materials are in short supply or if they want to hasten the study of the diseases.
   2. Securing a veterinarian as a resource person or taking a field trip to the veterinarian's office.
   3. Securing a resource person such as a hog producer who has had to deal with a disease problem recently or one who is known to have an excellent disease prevention program.
   4. Listing the books, bulletins, films, film strips, etc., to be used.
B. Resources for problem area #2
   1. An exhibit of swine parasites prepared by the class.
   2. A list of the references and teaching aids.

## PLAN NUMBER FOUR (A PROBLEM AREA PLAN)[3]

**Part I. Pre-planning Analysis**

A. Enterprise: Marketing agricultural commodities
B. Problem area: Using the futures market
C. Analysis of teaching situation:
   1. Class is made up of seniors who have had basic agricultural economics.
   2. All students work in agribusinesses.
   3. Prior study was just concluded on marketing alternatives.
D. Teaching / learning objectives: To develop the ability to:
   1. Explain how the futures market functions.
   2. Name at least 10 commodities bought and sold on the futures market.
   3. Describe how farmers can use the futures market.
   4. Describe how processors can use the futures market.
   5. Describe how speculators use the futures market.
   6. Define the terms, *bear, bull, short, long, maximum daily fluctuation, leverage, margin, hedger, speculator,* and *commission.*
   7. Name the locations of futures markets and brokers.

**Part II. The Teaching Plan**

A. Enterprise or activity: Marketing agricultural commodities
B. Problem area: Using the futures markets
C. Interest approach or motivation step:
   1. Have students play the card game, Pit.
   2. Have students buy or sell contracts; offer rewards to best speculators.
   3. Define futures market.
D. Anticipated group objectives:
   (Lead question: How can the futures market be used by farmers, agribusiness people, and speculators?)
   1. To make a greater profit on each crop.
   2. To avoid or minimize the risk of loss.
   3. To guarantee a selling price.
   4. To use financial resources effectively.
E. Anticipated problems and concerns of the students:
   (Lead question: If you are going to use the futures market, what are some things you will need to know?)
   1. How can I use the futures market?
   2. What can I buy and sell on the futures market?

---

[3]Adapted from a lesson plan developed by Dr. John Hillison.

## REVIEWING SOME ILLUSTRATIVE PLANS

3. How can I buy and sell on the futures market?
4. Who can use the futures market?
5. Where are the futures markets and brokers located?
6. What terms do I need to understand?

F. Steps in solving the problems. (Have the students select the problems from the list. They might be covered in the following order.)

NOTE: Many times teachers find it easier to arrange their lesson plan in a "T" arrangement as follows. This format permits the teaching resources to be aligned with the content to be covered.

| Source | Problems and Solutions |
|---|---|
| Teacher | **Problem 5.**<br>Where are the futures markets and brokers located?<br>  Chicago<br>  New York<br>  Minneapolis<br>  Kansas City<br>  Winnipeg |
| *Economics: Applications to Agriculture and Agribusiness,* p. 302. | **Problem 2.**<br>What can I buy and sell on the futures market?<br><br>  Lumber       Hams<br>  Potatoes     Live hogs<br>  Live cattle  Eggs<br>  Feeder cattle Butter<br>  Boneless beef Milo<br>  Pork bellies Turkeys |
| *Economics: Applications to Agriculture and Agribusiness,* pp. 301-303. | **Problem 1.**<br>How can I use the futures market?<br>  To hedge = To guarantee or protect the prices at which they will buy or sell listed commodities for cash some weeks or months in the future.<br>  To speculate = Use risk capital (money) to try to take advantage |

## TEACHING AGRICULTURE THROUGH PROBLEM SOLVING

of price fluctuations in the future market.

(A speculator puts no money in a parking meter and takes a chance on getting a ticket. A hedger puts a dime in the parking meter and knows there will be no ticket.)

Sample Problem

Example of farmer hedging corn
Sell + buy = closed contract
June sells December corn on
futures market for $2.50/bushel

| December buys corn on futures market for $2.40 | Cash price = $2.40 |
| --- | --- |
| profit = 10¢/bushel | $2.40 cash <br> + .10 profit <br> $2.50 |

June sells December corn on
futures market for $2.50/bushel

| December buys corn on futures market for $2.65/bushel | Cash price = $2.65 |
| --- | --- |
|  | $2.65 |
| loss = 15¢/bushel | − .15 loss <br> $2.50 |

*Agriculture Marketing: Systems, Coordination, Cash and Future Prices,* pp. 215-222

(When the farmer hedged, a guaranteed price of $2.50 was locked in.)

The *bull* thinks prices are going up.

The *bear* thinks prices are going down.

Discuss volume, spread, delivery period, etc.

Teacher

**Problem 3.**

How can I buy and sell on the futures market?

An order is given to a member firm. The member firm tells one of its brokers to either buy or sell for you.

You must pay a commission between $30 and $45 for the transaction. You must put up a margin or down payment.

REVIEWING SOME ILLUSTRATIVE PLANS 119

Teacher

**Problem 4.**

Who can use the futures market?
Anyone who has the money for a commission and margin.

The processor takes an opposite viewpoint to that of the producer.

A baker needs wheat in March.

June buys March wheat for $2.75/bushel

March sells wheat for $3.00/bushel

profit = 25¢

Cash price = $3.00
− .25
actual cost = $2.75

Teacher

**Problem 6.**

What terms do I need to understand?

*Short* = A trading position with more sales than purchases.

*Long* = A trading position with more purchases than sales.

*Maximum daily fluctuation* = The most a commodity can rise or fall before trading is stopped.

*Leverage* = The ability to control large amounts of value with relatively little cash.

G. Evaluation and application:
  1. Ask the students:
     a. What is the futures market?
     b. How can the futures market be used?
     c. What is the difference between a speculator and a hedger? What is the difference between a bull and a bear?
     d. What effect does the futures market have on price cycles?
     e. How can a farmer hedge live cattle on the futures market? Use specific prices and delivery months.
     f. How would you explain the futures market to someone who had never heard of it?

2. Have the students discuss what they have learned with their parents and employers. If their project sales are large enough, the students may be able to hedge on the futures market. They will be able to read the futures trends in the daily newspaper.

**Part III. Teaching Resources.**
1. Ewell P. Roy, Floyd L. Corty, and Gene D. Sullivan. *Economics: Applications to Agriculture and Agribusiness*, 2nd ed. Danville, Illinois: The Interstate Printers & Publishers, Inc., 1975.
2. Wayne D. Purcell. *Agriculture Marketing: Systems, Coordination, Cash and Future Prices*. Reston, Virginia: Reston Publishing Company, 1979.

## PLANS IN AGRICULTURAL MECHANICS

A problem-solving approach in teaching can be used also in providing instruction in the area of agricultural mechanics. Much of the instruction takes place in connection with the units and problem areas on housing, equipment, feeds, and the adjustment of machinery. However, most of the agricultural mechanics instruction takes place in connection with units and problem areas which are "lab" oriented. Some examples of these kinds of units and problem areas are:

Tractors, trucks, and engines
Caring for electric motors
Advanced agricultural welding
Agricultural machinery
Wiring agricultural buildings

Even though these kinds of units and problem areas are taught in the agriculture lab with the emphasis on the development of skills, there is always a certain amount of group and individual planning and related instruction which should take place in the classroom before any move is made in the direction of the lab itself. Good lab classes have their origin in the classroom. In the final analysis, the understandings developed are probably more important than the skills.

Perhaps the most difficult decisions for the teacher to make in relation to instruction in agricultural mechanics are those relating to the giving of demonstrations and those relating to when the

## REVIEWING SOME ILLUSTRATIVE PLANS

class is ready to start work in the lab. The skill demonstrations, including a minimum of student practice on the skills, can be scheduled as a part of each day's instruction. With regard to the actual initiation of student work on individual and group projects, it is well to keep the following in mind:

1. The students must be prepared well enough in the classroom to permit them to work in the lab for the day without further instruction by the teacher. The teacher simply does not have enough time in a 50-minute period or in an 80-minute period to spend much time with each student in a class of 15 to 20 eager youths.
2. Students need to be taught how to plan, how to order materials, how to select materials to be ordered, and a host of other things *before* they can be expected to have all needed materials in the school lab and be ready for the beginning of construction work. It simply does not make sense to say to the students, "We start in the lab on Monday if you have your materials here."
3. Poor lab management, lack of control in lab classes, and failure to abide by safety rules are a direct result of a failure to provide the necessary instruction before permitting work in the lab to begin. (See Chapter 11 for helpful suggestions.)
4. It is easier to motivate students to work hard in the classroom *before* starting an actual construction project than it is to wait until they are brought back to the classroom after having been exposed to the pleasure of lab work.
5. Each day of actual lab work for each class should begin in the classroom to review what happened in the lab the day before and discuss any problems that occurred there and to see if everyone will be able to work in the lab without help from the teacher.
6. Only by providing adequate pre-project construction instruction can the teacher be free for the task to be performed when a class is in the lab—constant supervision of the work of all of the students from vantage points near the centers of activity. If the class is very large, the teacher can afford only helpful hints and suggestions if the supervision is to head off serious mistakes and unsafe activities. Detailed instruction for one or two students is not compatible with the responsibilities for supervising the entire class.

INSTRUCTION IN LAB SKILLS SHOULD PRECEDE
THE NEED FOR THE SKILL.

## PLAN NUMBER FIVE (A PROBLEM AREA PLAN)

**Part I. Pre-planning Analysis**

    A. Enterprise or activity: Agricultural mechanics
    B. Problem area: Beginning agricultural carpentry
    C. Analysis of the teaching situation:
       1. Freshman class.
       2. No previous instruction in carpentry.
       3. Have considered housing and equipment in relation to:
          a. Raising the dairy calf.
          b. Starting plants in cold frames.
          c. Brooding and rearing chicks.
          d. Feeding wildlife.
       4. Just completed care and repair of hand tools.
    D. Teaching / learning objectives: To develop the ability of the

# REVIEWING SOME ILLUSTRATIVE PLANS

students to construct and repair equipment and housing required for their supervised agricultural experience programs:
1. To make and follow working plans.
2. To make out a bill of materials.
3. To use carpentry tools as required for a project.
4. To determine costs.

**Part II. The Teaching Plan.**
  A. Enterprise or activity: Agricultural mechanics
  B. Problem area: Beginning carpentry
  C. Interest approach (discussion questions):
(The assumption is made here that decisions have been reached in previous units as to the kind and amount of equipment needed for the supervised agricultural experience programs and that the teacher has noted other carpentry needs of the students in a record of supervisory visits. Naturally, some equipment already will have been provided for the supervised agricultural programs which have been started and this will have to be taken into account. It is also true that very little motivation is ordinarily required for lab instruction. However, the interest approach is still needed to serve the other purposes listed earlier.)
    1. What agricultural occupations require a knowledge of carpentry?
    2. What are some of the carpentry jobs that are necessary to the agricultural industry?
    3. How do you go about doing these jobs?
    4. How can you determine whether a carpentry job has been done well?
  D. Anticipated group objectives:
(Lead questions: Why should we be able to do some carpentry work? What were we thinking about when we considered equipment and housing in relation to dairy animals, poultry, plants, and wildlife?)
    1. We have to be able to do carpentry work because it costs too much to buy everything or to hire a carpenter, and because some needed equipment cannot be purchased ready-made.
    2. We need good housing and equipment in order to get the best results.
    3. If housing and equipment are not kept in good condition, they will not last long.
    4. The small buildings or projects we sell must be well made in order to attract customers.

E. Anticipated problems and concerns of students:
 1. *Stage 1.* Put a list of the things each student plans to make or repair on the board. Reference should be made to the planning done in relation to the units on raising the dairy calf, starting plants, brooding and rearing chicks, and feeding wildlife. Knowledge of the home situation is important—especially for those students who seem to have difficulty in deciding what to do.
 2. *Stage 2.* (Lead question: Now let's make a list of the things we have to know and be able to do in order to make the items of equipment we have just listed.)
    a. How to determine what materials to use.
    b. How to make a working plan.
    c. How to prepare a bill of materials.
    d. How to paint it.
    e. How to use the following tools:
       (1) Power saw.
       (2) Carpenter's square.
       (3) Level.
    f. How to place a value on the finished product.
    g. How to lay out rafters.
F. Steps in solving the problems:
 1. Select a problem.
 2. Discuss to see what is and is not known.
 3. Conduct class study on what is not known.
 4. Discussion and conclusions.
 5. Repeat first four steps until all problems are solved.
    (In this case, the teacher should help the students see that a particular order should be followed in solving the problems. Many of the questions which will be raised will have to be solved, in part, through demonstrations and by practice in the lab.
G. Evaluation and application:
 1. Grades on quality of planning and work in the lab.
 2. Test, if desired.
 3. Grades on work done by students in the home situation and on the job.

**Part III. Teaching Resources:**

(References, teaching aids, and teaching techniques)
 1. List of demonstrations.
 2. Construction of projects.
 3. List of references and teaching aids.

REVIEWING SOME ILLUSTRATIVE PLANS 125

## PLAN NUMBER SIX (A UNIT PLAN)

**Part I. Pre-planning Analysis**
   A. Enterprise or activity: Agricultural mechanics
   B. Unit: Electricity in agriculture
   C. Problem areas:
      1. Wiring agricultural buildings
      2. Caring for electric motors
      3. Selecting electric motors
      4. Using electrical controls
   D. Analysis of the teaching situation for the unit:
      1. Junior class.
      2. Class has had instruction in basic and advanced agricultural science.
      3. All homes of class members are supplied with electric power.
      4. Some new wiring and a great deal of electrical maintenance work are done on farms, in homes, and in nonfarm agricultural businesses.
   E. Teaching / learning objectives: To develop the ability of the students to use and to maintain electrical power systems in agriculture:
      1. To appreciate the uses of electricity.
      2. To understand where electricity comes from and how it is conducted.
      3. To diagram wiring projects and determine what materials are needed for a wiring project.
      4. To apply the electrical wiring code to wiring projects.
      5. To maintain electric motors.
      6. To use electricity with safety.

**Part II. The Teaching Plan**
   A. Enterprise or activity: Agricultural mechanics
   B. Unit: Electricity in agriculture
   C. Problem areas:
      1. Wiring agricultural buildings
      2. Caring for electric motors
      3. Selecting electric motors
      4. Using electric controls
   D. Interest approach for the unit:
      1. What are the different ways in which we use electricity in agriculture?
      2. Where does your electricity come from? How does it get to where you want to use it?

3. How many electric motors do you have in your home?
4. How many different kinds of electrical controls do you know about?
5. In what agricultural occupations would you need the knowledge and skills of electrical work?
6. What kinds of electrical work have been done at home?
7. How can you determine whether electrical work has been done well?

E. Anticipated group objectives for the unit:
(Lead question: Why should we be able to work well with electricity in agriculture?)
1. So we can obtain jobs which require knowledge and ability with electricity.
2. So we can do our own electrical repair work.
3. So we can make electricity do more work for us.
4. So we can use electricity with safety.

F. Study of problem areas:
1. Problem area #1. Wiring agricultural buildings.
   a. Discuss group objectives which apply.
   (Lead questions: Which of the unit objectives apply to this problem area? Are there additional objectives we should recognize?)
   b. Anticipated problems and concerns of the students:
   (1) *Stage 1.* List jobs for each student.
   (Lead question: We have listed on the board many of the wiring jobs that have been done around your home. Now let's make a list of the jobs that need doing right now at your homes or places of work—jobs that you might do.)
   (2) *Stage 2.* Problems. (Lead question: What do we need to know and be able to do in order to do the jobs you have listed?)
   (An alternative approach is to take one job listed by the students which involves all of the objectives of the teacher and ask the students what they need to know and be able to do for this one job. Then follow the job through to completion.)
   (a) What is electricity?
   (b) What kind of wire should we use?
   (c) What materials and fixtures should we use?
   (d) How do we get the electricity to where we want it?
   (e) How do we diagram a wiring project?
   (f) How do we make various connections?
   (g) How do we know whether it is safe?

(h) How do we locate the cause of trouble if something goes wrong?
(i) How do we read a meter?
(j) How do we prepare a bill of materials?
(k) How do we use the National Electrical Code?
(l) What tools and equipment do we need?
   c. Steps in solving the problems:
      (1) Lead the students in the selection of a problem.
      (2) Lead the discussion to find out what is and is not known.
      (3) Conduct class study on what is not known.
      (4) Lead the discussion to develop conclusions.
      (5) Have the students select another problem and repeat the process.
   d. Approved practices.
      (Note that some problems require demonstrations. These may be conducted in the lab, followed by practice for the students.)
 2. Problem area #2: Caring for electric motors
    (Repeat the steps taken for the first problem area for all remaining problem areas.)
G. Evaluation and application for the unit:
 1. Plans for projects to be done at home or elsewhere.
 2. Evaluation of shop exercises.
 3. Paper and pencil test.
 4. Instruction and supervision for projects.

**Part III. Teaching Resources**

(References, teaching aids, and teaching techniques for each problem area)
 1. List of demonstrations to be given, when they should be given, and in connection with which problems.
 2. Wiring jobs or exercises to be performed after classroom study and before projects are undertaken at home or elsewhere.
 3. National Electrical Code.

## PLAN NUMBER SEVEN (A PROBLEM AREA PLAN)

**Part I. Pre-planning Analysis**

A. Enterprise or activity: Orientation and guidance
B. Analysis of the teaching situation:
 1. Class made up of freshmen in latter part of the year.
 2. All of the class are minors.

3. A very small portion of the group has a definite occupational goal in mind.
4. Some of the group are coming to high school mainly because of parental pressure.
5. Most of the class will not be going on to higher education beyond high school.
6. On the average, one or two students out of the class will be going on to college.
7. The group has had no previous organized instruction on this particular problem area.

C. Teaching / learning objectives: To develop the ability to:
1. Use the help and services provided by the school.
2. Choose wisely from the courses and study programs offered.
3. Understand the values of the activities and organizations of the school.
4. Check on their general study habits with the aim of trying to improve them.
5. Examine personal goals regarding high school work.

**Part II. The Teaching Plan**

A. Enterprise or activity: Orientation and guidance
B. Problem area: Getting the most out of school
C. Interest approach (discussion questions):
1. How are you getting along in school?
2. Are you really getting what you want out of school?
3. What courses are you taking?
4. What grades are you making in them?
5. How are you using your free time?

D. Anticipated group objectives:
(Lead question: Why is it important that we "get the most out of school"?)
1. To get along better with our parents.
2. To be able to get better grades so we can make the honor roll.
3. To get a high school diploma.
4. To keep up eligibility for athletics.
5. To be able to earn a State Farmer degree.
6. To get started in farming.
7. To qualify for college entrance.
8. To get a better job.

E. Anticipated problems and concerns of the students:
(Lead question: What things do you need to know about yourself and the school in order to get the most out of school?)

*REVIEWING SOME ILLUSTRATIVE PLANS* 129

    1. Why do I have to go to school until I'm 16 when I'm not interested in school?
    2. What can I do if I don't like some of my teachers this year?
    3. What courses can I take next year?
    4. What can I do when my job takes so much time that I don't have time to study outside school?
    5. How do I know what courses are best for me to take?
    6. What can I do when my science teacher gives such long assignments that I don't have time to get my lessons in study hall?
    7. What good is school for me when I want to join the Navy as soon as I'm old enough?
    8. Which is better, to make high grades or to belong to more organizations?
    9. I plan to go on to college and study to be a veterinarian. How can school help me to find out more about it?
    10. How can we have more fun out of going to school?
    11. Dad says that I've got to stay in school until I graduate. What can I do to change his mind?

F. Steps in solving the problems:
(Most teachers no longer list these steps once they have become accustomed to them.)

G. Evaluation and application:
    1. Check over students' notebooks.
    2. Observe closely for change in behavior of students in regard to their study habits.
    3. Watch for indications of improved relationships between students and teacher.
    4. Observe for better understanding of the real purposes of school and all school activities.
    5. Look for evidence of more thoughtful and serious choices of courses and study programs.
    6. Develop a list of accepted practices—"Tips on How to Study."

## Part III. Teaching Resources

A. References, teaching aids, and teaching techniques:
    1. High school handbook.
    2. School schedule of courses.
    3. Lists of college entrance requirements.
    4. College catalogues.
    5. John C. Crystal and Richard N. Bolles. *Where Do I Go from Here with My Life?* Berkeley, California: Ten Speed Press, 1974.
    6. Dean L. Hummel and Carl McDaniels. *How to Help Your*

*Child Plan a Career*. Washington, D.C.: Acropolis Books Ltd., 1979.
7. Lorne Kelsey et al. *Eyes to the Future*. Pittsburgh, Pennsylvania: Stanwix House, Inc., 1973.
8. Principal talks to group about special services provided by the school.
9. Panel discussion by four students of the group on topic "How to Improve Study Habits."
10. Each member keeps a careful record of how time was spent through a typical day.
11. Coach meets with class to discuss problems raised about athletics.
12. Student leaders of some of the school organizations meet with class.

## TEACHING "NEW DEVELOPMENTS" UNITS

Many teachers find it difficult to teach the units on new developments. Actually, the application of the procedure to new developments units approximates the application to the disease and insect or parasite units. The suggestions which follow represent but one of the several ways a new developments unit could be taught.

1. The interest approach should relate to the entire enterprise with the emphasis being placed on changes in practice since the students last studied any unit or problem area on the enterprise.
2. The group objectives should be determined for the enterprise and should reflect any increased efficiency and production since the students last studied any unit or problem area on the enterprise.
3. The problems and concerns should be developed in two stages. The first stage would be the identification and listing of any new developments the students and the teacher wished to consider. The problem area titles could be used as a rough guide to make certain that consideration is given to all parts of the enterprise. The list of new developments would be considered as the problems to be solved. The second stage would be the development of a list of guide questions for the study of the new developments. Following the development of the guide questions, the teacher would take up the new developments one at a time with the entire class, or the teacher could assign each

new development to an individual student (or a small group of students) who would study about the new development and report the findings to the class.

## ADJUSTMENTS FOR AGRIBUSINESS OCCUPATIONS INSTRUCTORS

In a placement-employment program, students are placed in training stations in agri-related businesses. This practice has added a new dimension to the problems faced by the teacher in organizing classroom instruction.

These placement-employment programs have created at least two new situations in agricultural departments. One occurs if only one or two students elect to make use of the opportunity to secure agricultural experience through the program. These students will be enrolled in vocational agriculture classes in which the agricultural content will be organized as it always has been organized. The other situation occurs if there are enough students interested in securing agricultural experience through the program to warrant enrolling them in a separate class.

Following is a discussion of each situation.

### Agricultural Departments Having Only a Few Students with Placement-Employment Programs

With only one or two students in the class making use of placement-employment programs, the teacher of agriculture will probably use group instruction techniques for the major portion of the class time. This means that the course content will be organized much as it has always been organized and that the content will be planned with the majority of the students in mind. The majority of the students will undoubtedly have programs of the kind made familiar to the field generally over the years in which vocational agriculture has been taught.

Some adjustments will need to be made to provide the placement-employment students with needed instruction. A small part of this adjustment may be in the form of adding some group instructional units which meet the needs of the placement-employment students more than they do the needs of the students with the usual farming programs. The major part of the adjustment, however, will probably be in the form of making more use of individual and small group instruction techniques.

Using the teaching plan as a study guide can be very useful in planning for individual and small group instruction. It can serve to give direction to the work of the placement-employment students when the teacher wishes to have them work individually or in a group separate from the students having farming programs. The students can follow the organization of the teaching plan as they study in class and as they obtain information about the businesses where they work.

### Agricultural Departments Having Separate Classes for Placement-Employment Programs

The problems of adjusting instructional programs are less difficult in situations in which enrollments are large enough to require setting up separate classes for those who are in the placement-employment program. In these situations, the basic approach to teaching is the problem-solving approach presented in this publication. The teacher will need to decide on the technical agriculture content for the class, since this content will vary greatly from community to community and from state to state. The unit and problem area list in Appendix B provides suggestions for determining the content for separate classes for students having placement-employment programs.

Teachers may have some problems in organizing the work for these kinds of classes. Planning will need to be done in terms of group instruction for content common to all in technical agriculture and in the business content. In addition, plans will have to be made for small group and individual instruction for both the technical agriculture content and the business content. A typical week of classes may follow an outline of activities similar to the following:

1. A discussion of problems brought up by the students and problems identified by the teacher as on-the-job supervision and instruction.
2. The giving of reports by individual students on their work and on their individual study of the products, services, and other aspects of their training stations.
3. Individual and small-group study by the students. Some

teachers may wish to meet individual needs entirely through homework assignments. This would leave all of the class time for large-group and small-group work.
4. Full-class instruction on technical agriculture content or on agriculture business content.

## Obtaining Information

Some teachers are finding that it is difficult for the students to find time to obtain certain of the items of information about their training stations. The students are actually employed for wages, and they seem to be reluctant to take time off to study the agricultural businesses as they must in order to profit from the experience as much as they should.

Part of the problem of obtaining information can be solved by the simple expedient of having the students take some of the literature related to the products and services of the businesses home to study or to the classroom to study. For information which can be obtained only during working hours at the business itself, however, it appears necessary to arrange for each student to have an hour or two in each week scheduled for the purpose of studying the business. These study hours should be scheduled in addition to the working hours and no pay should be expected for them. In some businesses in which the work is lighter at certain periods of each week or month, the study hours can be scheduled to coincide with these periods of low business activity. This would also make other personnel in the business more available for answering questions.

Not all of the information students should obtain can be obtained at once. Some of the units of study need to be introduced early in the school year with the expectation that the students will then be obtaining that kind of information, recording it in the notebook, and reporting on it to the class throughout the entire school year.

The most important unit[4] in this respect is the one containing the problem areas on getting acquainted with the products and services of the agricultural business. These two problem areas are taught early in the school year with each student using a product or service from the training station as the basis for study. From

---

[4]See Appendix B.

the time the classroom instruction has been completed for learning about products and services, the students should study additional products and services individually. Monthly or biweekly reports on new products or services should be made to the class to encourage continuing individual study on products and services and to broaden the base of agricultural knowledge of the entire class.

It is obvious from the preceding discussion that the students should expect to carry their notebooks with them when they go to work and during the periods of time to be devoted exclusively to a study of the businesses in which they are placed.

Neither students nor teachers should expect the task of obtaining information about the businesses to be easy. There are no shortcuts to learning. When teachers and students begin to take shortcuts, they are at the same time beginning to shortchange the students' educational programs and individual development.

Some businesses may be unwilling to provide some of the information desired. If this is true, the teacher will have to decide whether the business is a good place in which to place students for agricultural experience.

If it is decided that the business is a good situation, the teacher will then need to supplement the information obtained by providing references and other sources for those kinds of information made unavailable by the business. Field trips and resource persons can be used effectively to provide students with additional information about businesses.

# CHAPTER 6

# Taking the Plan Into the Classroom

A PLAN IS OF NO VALUE unless it can be followed. In this case, the plan must reflect the natural order of events to be followed when the teacher goes before the class.

This chapter is an attempt to provide a glimpse of the way in which the teaching plan would be used in actually teaching the class. The procedure is outlined step by step as it would normally develop. It is the task of the teacher to lead or guide the students through the development of the plan. Skillful questioning and discussion-leading techniques will be needed to avoid long and distracting detours while still maintaining an atmosphere of joint planning and working together toward common goals.

The discussion does not show all parts of the teaching plan as being developed with the students. This is the teacher's plan, not the student's plan. This discussion does, however, show all parts of the plan where students play a major role in development or in responding to the teacher's initiatives.

### WHAT THE TEACHER DOES IN THE CLASSROOM

1. *Write the enterprise, unit, and problem areas on the board.*—This should be accompanied by a brief orientation as to the origin of the unit or problem area and a statement to the effect that this will be the area of study for the next few days. This helps to set the stage for what follows.

    If the content segment is in mechanics, the students should be informed about how the classroom and lab work will be coordinated and about what lab projects or other application activities are expected. In the case of any laboratory activity, the foundation for successful performance is built through careful classroom instruction before start-

**LET THE STUDENTS TELL HOW THEY ARE DOING THE JOB.**

ing the laboratory activity and by careful daily supervision and review.

2. *Develop the interest approach.*—The purposes of the interest approach have already been discussed. Care must be taken at this stage to avoid being drawn into a problem-solving situation and to avoid letting the discussion drag on to the point at which the students are not contributing their experiences spontaneously. This can be controlled quite easily by asking the next question in the interest approach and by not yielding to the temptation to "give" the students information, since differences will form the basis for many of the student problems and concerns. The teacher must remember that it is not important at the time of the interest approach that the statements made by the students be cor-

rect from a technical point of view. It is important that the students recognize that they differ among themselves with regard to various aspects of the unit or problem area. A sharing of experiences at this point will provide those students having no knowledge in the area with enough background to permit active participation in the teaching / learning process.

3. *Develop the group objectives.*—The development of student (group) objectives should grow naturally from the interest approach. The lead question will keep the discussion moving in the proper direction. The key question to keep in mind is, "Why is it important to become familiar with (this area)?" If the objectives were developed as the interest approach was developed, it should take very little time to summarize them and list them on the board.

4. *If this is a unit plan, select the first problem area to be considered, and relate the problem area to the objectives. If it is a problem area plan, move directly to step 5.*

    For a unit, it is important that the students relate each of the problem areas in the unit to the student (group) objectives for the unit. It is well for the teacher to be reminded often that the group objectives help make the instruction meaningful to the students. Thus, the teacher should ask the students to point out the specific objectives to which a study of the problem area selected should contribute.

5. *List the problems and concerns of the students.*—Sometimes the students begin to state problems without any encouragement from the teacher. In most cases, however, the teacher will have to lead the way with suitable questions. The teacher will continue to ask leading questions until the class has listed problems which will include all of the things which need to be considered as indicated by the teaching plan.

    It is important that the teacher record the problems given by the students in question form and by using the words of the students. Rephrasing by the teacher may discourage students from contributing, or it may result in a problem which does not carry the meaning intended by the student.

It has been noted that the students are not asked to "list their problems." Instead, they are asked to list *the things that they know how to do and things that they need to know in order to accomplish their objectives with regard to the area under consideration.* Many times, a particular item listed does not become a problem for the students until the initial discussion during the problem-solving stage when they realize for the first time either that they do not know the answer or that their ideas differ from those of other members of the class.

Listing all of the problems and concerns before starting to solve any of them gives both the teacher and the students an overview of the entire problem area. This will help avoid overlapping and duplication and will assure all of the students that their particular problems and concerns will be considered.

6. *Follow the steps for solving the problems.*

a. *Select the problem(s).*—At this point, the teacher may learn from the students with which problem, of those listed, they wish to begin. If it is desirable, the teacher may lead the class to determine the best order in which to consider all of the problems before any are solved.

There is a matter of common judgment on the part of the teacher which must enter in at this point. In listing the problems and concerns of the students, some of the problems may have been of a minor nature requiring only a minute to locate the solution. Other problems may be listed twice in different words, or they may be duplicated in some other way such as being included in a problem more broadly stated. Still other problems may be much too broad in scope for the class to consider without further subdivision. It is the teacher's task and responsibility to lead the class to consider minor problems as a part of broader problems; to lead the class to consider two or more problems at the same time, if this seems warranted; or to lead the class to further subdivide problems which are too broad. Most of these difficulties can be avoided with proper care and attention at the time of the listing of the problems.

**TAKING THE PLAN INTO THE CLASSROOM** 141

IF THE STUDENTS CAN SEE ALL OF THE PROBLEMS TOGETHER, THEY WILL NOT BE GOING OFF ON TANGENTS IN ORDER TO MAKE SURE THAT THEIR PARTICULAR PROBLEMS ARE ANSWERED.

b. *Lead discussion to find out what the students do and do not know.*—After the problem has been selected, it should be discussed in order to determine what the students do and do not know about the problem. This discussion may result in a series of questions which can be listed on the board as a guide for the students to follow in their study.

In some cases, the students may already know and understand the solution, and conclusions can be drawn without resorting to supervised study.

c. *Conduct supervised study.*—Following the discussion, the students should use the references and other materials to find the solution to the problem or to that part of the problem for which they did not have a correct solution. If the students do not have a good understanding of the solution to the problem, there is a need for study even though the solution is correct.

Not all problems can be solved through the use of literature. In some cases field trips, resource persons (including the teacher), or teaching aids may be required. In most cases there will be supervised study. During supervised study the teacher should work closely with the individual students to give help where needed and to keep a close check on the progress being made.

Some teachers find it difficult to begin a supervised study period gracefully or smoothly. A simple statement such as "Let's see what we can find out about these things" or "Where can we find out about these things we need to check on?" will usually be satisfactory.

Some problems, especially when laboratory activities are involved, require the teacher to provide for demonstra-

LISTING OF APPROVED PRACTICES ENCOURAGES
APPLICATION ON THE HOME FARM.

tions. Demonstrations should follow student study and discussion whether they are only to be observed or serve as instructional steps in preparing students to perform the activity being demonstrated.

Some teachers prefer to complete all of the classroom instruction and student planning for projects prior to beginning laboratory demonstrations (and activity) in order to provide the maximum time possible within the class period for those activities. Whether demonstrations are provided throughout the classroom study time or during the laboratory work time is not critical. It is critical that the students learn as much as possible about the laboratory activities before starting to do the laboratory work.

d. *Lead final discussion.*—The last step in the solving of a problem is, of course, the final discussion and the drawing of conclusions. By working with the students, the teacher can easily determine when study should cease and when the final discussion should begin. The length of the supervised study period is automatically determined by the students themselves. As soon as most of them have arrived at their individual solutions, it is time for the teacher to initiate the concluding discussion with a statement such as "Most of you seem to be finished. What did you find out?" Those students who are very slow and do not finish will contribute what they can and get the rest through the discussion. The students who are quick to arrive at their solutions can be encouraged to look more deeply into the "why" of the solution in additional references, to give assistance to the slower students, or to place their solutions on the board as a basis for the concluding discussion.

7. *Repeat the procedure indicated in step 6 until all of the problems have been solved.*

8. *Review the entire problem area studied by leading the students in the development of a single list of all approved practices decided upon.*

While the instructional method proposed here is problems oriented, it is still appropriate to identify any principles which have been developed. As instruction is provided on other similar areas, the principles can again be pointed out to help the students appreciate how knowledge can be applied to solving problems in a variety of situations.

## CONDUCTING LABORATORY ACTIVITIES

Laboratory activities usually continue for some time beyond the ending of the class instruction which prepared the students for those activities. However, some group sessions are still needed.

At the start of each class, the teacher should meet the students in the classroom (or in the lab) to review what went on the day before, to make sure each student knows what is to be done that day, and to provide for any demonstrations that are needed. These daily group sessions help the teacher make the best use of any available time in working with each student in the class and contributing to the achievement of sound management of the class. See Chapter 11 for a more complete discussion on conducting laboratory activities.

# CHAPTER 7

# ▶ Avoiding Trouble Spots ◀

TEACHERS SHOULD NOT EXPECT that all will go perfectly with the very first attempt to use the teaching plan suggested in this publication. Each teacher must work to develop the ability along the lines required by the plan. A list of some of the more common "trouble spots" may be of some help to teachers in analyzing their difficulties in using this particular problem-solving approach to teaching.

1. *Unsuitable or poorly defined unit or problem area.*—This usually results in a lack of response from the students because of misunderstandings or in the listing of problems not related to the area being studied. The teacher will usually appear to be confused.

2. *Inadequate teacher analysis of the situation in relation to the unit or problem area.*—If teachers do not have a good understanding of the situation as it exists in relation to a particular unit or problem area, they may find themselves working at cross purposes with the students.

3. *Inadequate teaching / learning objectives.*—Inadequate teaching / learning objectives may be caused by a failure to list objectives prior to making plans, by not recognizing all of the specific objectives for the unit or problem area, or by failing to determine objectives in terms of the changes which the teacher wishes to bring about in the behavior of the students. This may result in a poor teaching plan and confusion in the teacher's mind as to the direction in which to lead the discussion and the thinking of the students.

4. *Omission of the interest approach, or allowing the interest approach to become a problem-solving situation.*—Failure in respect to the proper place and use of the interest approach usually results in a rather incomplete and unsatis-

POORLY DEFINED UNITS OR PROBLEM AREAS
CAN LEAD TO CONFUSION ON THE PART
OF THE TEACHER AND STUDENT

factory study of the unit or problem area. If the interest approach is not well developed, there will be a lack of enthusiasm on the part of the students, a lack of understanding of the problem area and how each problem is related to the entire picture, an inadequate use of source materials, and an uneven response because of the failure to provide a common background of experiences before analysis of the unit or problem area is started.

5. *Poor discussion-leading techniques.*—It is the teacher's responsibility to determine the direction in which the discussion will move and to ask the right kinds of questions to accomplish this movement. In addition to asking the right kinds of questions, the teacher must be able to realize at what point the discussion has served its purpose and then lead the class on to more fruitful activities.

Another problem is that the more advanced students sometimes provide the correct solution before the majority

# AVOIDING TROUBLE SPOTS

GET OUT ON THE FARMS AND STUDY
THE FARMING SITUATIONS.

of the class can be called upon. This can be controlled by giving all students an opportunity to respond, by asking for the *why* of the particular answer from another student, and by moving into supervised study as soon as it is evident that most of the students do not understand or that there are differences of opinion as to the correct solution. Much of the discussion-leading difficulty comes in the effort to determine what the students do and do not know about a particular problem. Only three or four responses may be needed to reveal the need for study to both the teacher and the students.

6. *Lack of orientation at the start and at the end of the class period.*—Students need to be "warmed up" to the work for the day by a brief review of what has been accomplished thus far. The teacher may summarize the work or may call

on the students to do so. If the work is stopped a few minutes before the end of the period for a brief review of the day's progress and some indication of where the next day's work will start, the "warming up" process will be much easier.

7. *Taking too many problems for study at one time.*—Taking too many problems to solve at one time may cause the slower students to get further behind than usual, or it may result in confused solutions because of a lack of understanding of how far to go with a particular problem. It also means a longer period of supervised study and longer periods of discussion. These may cause unrest on the part of the students. Taking one problem at a time provides for a variety of activity through alternating shorter periods of discussion and study, and it provides the opportunity

THE USE OF NOTEBOOKS HELPS THE TEACHER DETERMINE
WHERE HELP IS NEEDED MOST DURING
SUPERVISED STUDY.

needed to use those teaching techniques best suited to a particular problem.

8. *Not using notebooks properly.*—The use of notebooks by the students should be looked upon as a valuable teaching aid. When students write their solutions in their notebooks, it is easy for the teacher to determine who needs help, whether a particular student is studying, and when it is time to begin the concluding discussion. It also helps the student to learn by adding other ways in which to respond and by providing the means for self-evaluation of the ability to solve problems. The fact that many students do not keep their notebooks beyond the end of the school year is unimportant. The real value of the use of notebooks is in the extent to which it makes teaching more effective.

The student's notebook should contain the following items:
a. The name of the enterprise or activity.
b. The statement of the unit and/or problem area.
c. The lists of the student objectives, problems, and concerns.
d. The student's individual solution to each problem.
e. The corrections to the student's solution, if class discussion shows that corrections are needed.
f. The list of approved practices and principles identified.
g. The plans for the supervised experience programs.

Some teachers believe that the students should write only the class conclusions in the notebook in order to have a neat appearing notebook. This would, of course, make the notebook that much less valuable as an instrument for increasing the effectiveness of teaching. "To have a neat notebook" is not an important reason for the use of notebooks, although neatness should be encouraged.

9. *Omission of one or more steps in the procedure for solving the problems.*—Omitting any of the steps in the procedure for solving the problems may result in confusion through a lack of understanding, or in a lack of interest through being forced to study something which is already known. It is an important objective of education to teach students how to solve problems in a systematic manner.

10. *Inadequate source materials.*—Students cannot be expected to arrive at suitable solutions to problems if the resource materials are inadequate as to either quantity or quality.

11. *Failure to orient the students to a problem-solving approach.*—The use of a problem-solving approach in teaching places many new responsibilities on the students. It is unreasonable to expect them to adjust to the new procedures without adequate orientation. Furthermore, the teacher will need to use the problem-solving approach for several periods before students begin to feel comfortable with it.

12. *Taking the "teacher" out of "teacher-student" planning.*—Many of the failures of teachers with a problem-solving approach in teaching can be traced to the erroneous impression that placing more responsibility upon the students reduces the teacher's responsibilities. Actually, the teacher's responsibilities are much greater, although of a slightly different nature than they are with other methods of teaching.

13. *Failure to make complete plans.*—Some parts of the plan as outlined can be omitted after the teacher has developed the ability to use a problem-solving approach. The trouble starts when the teacher yields to such temptation too soon.

14. *Failure to provide for a variety of activities and procedures within the framework of a problem-solving approach.*—The use of a variety of techniques will make school more interesting for the teacher as well as for the students. This variety of activity can be provided in various ways. The following are a few illustrations of what can be done:
    a. Let the students help decide on some of the special events and activities.
    b. Set up tables to be filled in by the students as their solutions to certain kinds of problems. For example, when considering varieties of roses or the different breeds of cattle, the varieties or breeds can be listed down one side of the page, and the factors which the class thinks are important can be listed across the top of the page. This leaves a table to be filled in providing a slightly

## AVOIDING TROUBLE SPOTS 153

different situation than the problems presented in the statement form. A table can be set up for comparing shade trees, small engines, types of greenhouses, kinds of insurance, or other categories of items.

c. Plan for field trips with the students.
d. Let the students give certain demonstrations.
e. Use small-group work on some problems.
f. Let the students in the advanced classes lead the discussion occasionally.
g. Use film strips, charts, and other teaching aids.
h. Use "student panels" for certain problems.
i. Instead of the usual kinds of problem statements, use the insect, parasite, machine, or disease as the problem.

Refer to Chapter 4 for a complete listing of teaching techniques.

15. *A lack of understanding as to where teacher-student planning begins.*—Some of the confusion in the use of a problem-solving approach in teaching stems from a lack of understanding as to where and how teacher-student planning begins. Actually, teacher-student planning can begin with the planning of the entire course of study at the beginning of each school year. In this case, there would be a series of steps which should be followed, each with particular objectives in mind.

The first step would be to determine, with the students, the various enterprises, units, and activities which should be included in the year's work.

The second step would be to determine, with the students, what units and problem areas should be included for this particular year's work within each of the enterprises or activities previously decided upon and the time of the school year in which each unit and problem area should be considered.

The third step would be the teacher-student planning as presented in this publication in relation to the use of a problem-solving approach in teaching.

The first two steps involve what is known as either planning the course of study or planning the course calendar for the year. Most teachers perform this task them-

154    TEACHING AGRICULTURE THROUGH PROBLEM SOLVING

A VARIETY OF TECHNIQUES IS POSSIBLE WITHIN THE FRAMEWORK
OF THE PROBLEM-SOLVING PROCEDURE.

selves, using all available information, without the active assistance of the students. Some teachers do plan the course of study for the year with the students.

It is important here to understand the relationship of the first two steps to the third step. Each step serves to narrow the area of study to a more manageable unit. The final analysis comes with the breakdown of the problem area into subproblems for discussion and study.

16. *Moving along too slowly or too rapidly.*—Both the class and the teacher help set the pace when a problem-solving approach is used. If the teacher does not keep pace with the class, the students will soon become impatient and resort to undesirable side-play or lose interest entirely.

## AVOIDING TROUBLE SPOTS

17. *Failure to recognize a problem when it is suggested by a student.*—Many times students suggest problems by what they say, but the teacher fails to recognize the suggestions as such. For example, a student might suggest a lack of knowledge about what electricity is. This could easily be passed over if the teacher did not happen to be alert. The teacher could make use of this suggestion by saying, "Well, we certainly should know what electricity is, shouldn't we?" and then promptly write the statement on the board as a question.

    The teacher who anticipates the student problems and concerns will be much more apt to recognize problems when the students suggest them in what they say.

18. *Failure to adjust the teaching plan to local conditions.*—As was pointed out earlier, only the teacher in the school can plan properly for a particular class or group. There are too many factors in the local situation which must be taken into account. Thus, the presentation here is meant to provide only a sound framework and basic procedure which will be effective if the teacher adjusts and adapts planning to the local situation about which only the teacher can know.

19. *Failure of students to respond with the anticipated list of questions.*—It is too much to expect that all will go smoothly the first time a new procedure is attempted. One of the most trying situations is one in which the students fail to list any problems. If the students do not list problems as anticipated, the teacher should simply say, "Well, then, I have a few problems here which I think are important from having visited your homes." Then, the teacher should list the anticipated student problems and concerns on the board.

    At the end of the class, the teacher should carefully analyze all teaching procedures used to determine just why the students failed to respond.

20. *Failure to understand the proper relationship between unit and problem area.*—The unit concept is useful to help reduce the feeling of needless repetition which comes when two or more problem areas in the same enterprise are taught consecutively. The interest approach and setting of

objectives can be done once. However, continuing reference must be made to the objectives as they relate to each problem area and to each problem, especially when the instructional situation spans more than one day.

# CHAPTER 8

# ◆ A Minimum Plan ◆

MANY TEACHERS HAVE SAID "The plan looks fine, but it would take too long to go into all of that detail each time. What would be a minimum plan?"

It is true that the detail presented in this publication is partially due to the necessity for providing a complete explanation for the teacher who is totally unfamiliar with a problem-solving approach in teaching. It is just as true that shortcuts in planning for teaching often lead to "disorganized confusion" in the classroom.

However, since there is some justification for shortcuts for the teacher who knows the community well and who is experienced in the use of a problem-solving approach in teaching, the following outline is suggested as a guide for a minimum plan.

Two cautions bear repeating:
1. The minimum plan should be used only after the teacher has developed all of the abilities necessary for using a problem-solving approach in teaching.
2. The complete plan should be used whenever a problem area is being presented for the first time or whenever the teacher is not familiar with the local community situation.

### A MINIMUM PLAN

**I. Enterprise:**

(The need for this is self-explanatory.)

**II. Unit and Problem Areas or Content Segment:**

(The need for this is self-explanatory.)

**III. Teaching / Learning Objectives:**

This is the most difficult and important pre-planning step of the plan. How can the teacher make plans for guiding the students toward certain objectives if the teacher does not know what

**DON'T TAKE "SHORTCUTS" WITH TEACHING PLANS UNTIL YOU ARE SURE YOU KNOW WHAT YOU'RE DOING**

these objectives are? If the teacher carefully and thoughtfully prepares this step, making the rest of the plan will be relatively easy.

**IV. Teaching Procedures:**
   A. Interest approach:
   A few brief notes to indicate the general direction of development, any particular situations to which reference should be made, and the point of departure for the development of group objectives will probably suffice.
   B. Anticipated group objectives:
   It is particularly important that the group objectives be identified in advance if the situation is not recorded. This will help keep the teaching in line with local possibilities.
   C. Anticipated problems and concerns of the students: This list is needed to assist the teacher in recognizing the problems when they are raised, in getting problems stated which will encompass all of the teacher's objectives, and in getting problems stated which are suitable for study. A separate list should be prepared for each problem area.

## A MINIMUM PLAN

D. List of approved practices, test, and principles:
The list of approved practices should be carefully thought through in advance. It would be very easy for the teacher to overlook important practices when developing the list with the students. As pointed out earlier, constructing the test in advance will insure at least a minimum amount of attention to the development of a sound examination. It will also help the teacher test according to what the teacher is trying to accomplish rather than on the basis of incidental learnings. The test will be fairer to both the teacher and the students.

NOTE: Parts A, B, C, and D of the teaching procedure should be prepared in advance for still another reason. It is difficult enough under the best of conditions to do a thorough and masterful job of leading a group of enthusiastic young students in discussion toward a particular objective. Anything the teacher can do to be free from the nagging worry of missing some essential point, or wondering just what should be done next, should be done. That will leave the teacher's mind free to concentrate on teaching techniques, the comments and discussion of the class, and other factors which affect the quality of the classroom situation. There is no single factor which is more conducive to successful teaching than a good plan in which the teacher has faith.

E. References, teaching aids, and teaching techniques:
The need for this is self-explanatory.

## THE MINIMUM PLAN ILLUSTRATED

A Problem Area Plan

**I. Enterprise or Activity:**

Forage crops

**II. Problem Area:**

Improving and maintaining permanent pastures.

**III. Teaching / Learning Objectives:**

To develop the ability of the students:
A. To recognize when a pasture needs renovating.
B. To determine the fertilizer needs of a worn-out pasture.
C. To select the proper seed mixture.

162  TEACHING AGRICULTURE THROUGH PROBLEM SOLVING

   D. To prepare the pasture for reseeding.
   E. To describe the value of a good pasture.
**IV. Teaching Procedures:**
   A. Interest approach:
      1. Field trip to:
         a. Farm to see a good pasture.
         b. Farm to see a poor pasture.
      2. Discuss the two pastures from all angles—grasses and legumes present, length of grazing season, stock they will carry, etc.
      3. Raise the question of how much a good pasture should be worth.
   B. Anticipated group objectives:
      1. To provide pasture during the entire growing season.
      2. To have a pasture that will carry _____ per acre.

A LIST OF APPROVED PRACTICES SERVES AS A
VALUABLE SUMMARY.

## A MINIMUM PLAN

C. Anticipated problems and concerns of the students:
   1. What grasses and legumes make a good pasture?
   2. How should the pasture be prepared for seeding?
   3. When should the pasture be reseeded?
   4. What fertilizer should be used and how much?
   5. When and how should the fertilizer be applied?
   6. Where can the seed be purchased?
   7. How can it be determined whether the seed is good?
   8. What is a good pasture?
D. Evaluation and application:
   1. List of approved practices.
   2. Test (attach to plan).
   3. List of principles.

...AND HE DIDN'T EVEN HAVE A
TEST READY TO GIVE!

E. References, teaching aids, and teaching techniques:
1. Soil tests from pastures.
2. Examination of samples of seed.
3. Laboratory work on pasture plant and weed identification.
4. List of references.

## A Unit Plan

### I. Enterprise:
Landscaping

### II. Unit:
Establishing lawns and turf

### III. Problem Areas:
A. Selecting tools and equipment.
B. Preparing the seed bed.
C. Selecting the lawn or turf grass.
D. Seeding lawns.
E. Transplanting turf.
F. Caring for and maintaining tools and equipment.

### IV. Teaching / Learning Objectives:
To develop the ability of the students:
A. To select the proper tools and equipment for establishing lawns and turf.
B. To determine the best variety of grass for the situation.
C. To select the best make of tools and equipment.
D. To determine the best source of seed and turf.
E. To determine the best time to sow grass seed and to transplant turf.
F. To prepare the soil for planting.
G. To prepare the soil for transplanting turf.
H. To prepare the turf for transplanting.
I. To lay the turf.
J. To clean and lubricate the tools and equipment.
K. To sharpen those tools requiring sharp edges.
L. To repair tools and equipment.
M. To adjust tools and equipment.
N. To determine the plant food needs for grass and turf and to apply what is needed.
O. To apply needed herbicides and insecticides.
P. To plan for and apply the water needed by the plants.
Q. To determine the costs of establishing lawns and turf.

## A MINIMUM PLAN

V. **Teaching Procedures**
   A. Interest approach for unit:
      1. Ask students to describe how they maintain lawns and turf at home.
      2. Develop a list of occupations for which a knowledge of lawns and turf establishment is needed.
      3. Ask students how well persons in the local area perform these tasks.
   B. Anticipated group objectives:
      1. To prepare for a job.
      2. To keep the home grounds looking nice.
   C. Study of problem areas:
      1. Problem area #1: Selecting tools and equipment
         a. Relate to group objectives for unit.
         b. Problems and concerns:
            (1) What hand tools are needed for lawns? Turf?
            (2) What power tools are needed for lawns? Turf?
            (3) What special equipment is needed for lawns? Turf?
            (4) Where can the needed tools and equipment be purchased?
            (5) What makes of tools and equipment are best?
            (6) Should the tools and equipment be rented or purchased?
         c. Teaching resources
            (1) Carter – *Insects in Relation to Plant*
                Conover – *Grounds Maintenance Handbook*
                Leopold – *Water: A Primer*
                Musser – *Turf Management*
                Sprague – *Turf Management Handbook*
                Williams – *Nursery Crops and Landscape Design for Agribusiness Studies*
            (2) Aids:
                (a) Turf samples
                (b) Slides of grass varieties
            (3) Field trip to nursery
      2. Problem area #2: Preparing the seed bed
         (Repeat steps used for problem area #1 in teaching all other problem areas.)

VI. **Evaluation and Application**
   A. Tests
      1. Completion
      2. True and false
   B. Have students develop plan for improving home lawn.
   C. Summarize approved practices.
   D. Identify principles.

# CHAPTER 9

# Variations in Using Problem Solving

THIS PRESENTATION WOULD BE INCOMPLETE without some discussion of the many variations in the use of problem-solving techniques in teaching.

It is true that each teacher will use a problem-solving approach in a slightly different manner. Each teacher's personality will determine which approach that particular teacher will adapt for use in the classroom. The basic principles will, however, remain the same. A few of the variations in use are presented here for consideration. Each variation has been identified and defined through the observation of experienced teachers in their classrooms.

## VARIATION NUMBER ONE

One variation in the use of a problem-solving approach is that of combining the interest approach and the setting of the group objectives with the listing of each student problem or concern. This really amounts to having a separate interest approach for each problem which the teacher wishes to draw from the students.

The general procedure in using this variation is as follows:

1. Stimulate discussion of a specific aspect of a problem area according to the particular problem which the teacher wishes to draw from the students. This would involve a discussion of the experiences of the students and other factors which are a part of the interest approach as presented in this publication—but only in relation to the specific aspect of the problem area decided upon by the teacher.
2. Consider livestock or crop production records and goals as related to the specific aspect of the problem area under

PROBLEM-SOLVING PROCEDURES CAN BE ADAPTED
TO THE PERSONALITY OF THE TEACHER

discussion. This should result in the setting of objectives by the class.
3. Record the problem as stated by the teacher.
4. Repeat steps 1, 2, and 3 for the next problem which the teacher wishes to develop—and continue to repeat these steps until all of the problems for which the teacher had planned have been developed.

This variation has been used quite successfully. Its major weakness is that the initiative is kept in the hands of the teacher. This makes little provision for getting problems which are the result of a feeling of need on the part of the students. It also tends to limit discussion to a smaller number of students, since no general exchange of experiences has taken place prior to the stating of a particular problem.

## VARIATION NUMBER TWO

A second variation in the use of a problem-solving approach in teaching is that of solving each problem as it is raised.

Starting with the interest approach, the steps in this variation are as follows:
1. Stimulate discussion of the problem area.
2. Develop the group objectives.
3. Identify one problem which needs to be solved in order to achieve the group objectives.
4. Discuss the problem to determine what the students do and do not know.
5. Conduct supervised study or other teaching / learning activities.
6. Lead a discussion on what was learned, and develop a solution to the problem.
7. Identify a second problem, and repeat steps 2 through 6. Continue with this procedure until no more problems can be identified.
8. Bring study of the problem area to a conclusion by summarizing all of the solutions, developing lists of approved practices and principles, and evaluating student learning.

This particular approach has several weak points.
1. It gives the students no idea whatever of the total scope of the problem area.
2. It is difficult to organize for discussion and study.
3. There is a strong tendency to stray beyond the confines of the particular problem because the students do not know what is to follow or whether their particular interests will be considered.
4. Often discussion is abruptly limited because "we already have a problem to study."
5. Major points are easily overlooked. Since there is no overall planning from the start, it would be easy to think that particular points had been considered in relation to some problem already discussed.

## VARIATION NUMBER THREE

A third variation goes much further in the direction of overall planning than does the second. With this approach, each problem is thoroughly discussed when it is raised. This eliminates from further consideration those problems for which the students have correct solutions. Only those problems for which the students had no solutions are listed on the board to form the basis for supervised study and continued discussion.

Starting with the interest approach, the steps in this variation are as follows:
1. Stimulate discussion (interest approach).
2. Develop group objectives.
3. Identify a problem (or concern).
   a. Discuss the problem to determine whether it needs to be studied.
   b. If the solution is known, conclude the discussion. If the solution is not known, write the problem on the board for study.
4. Identify another problem.
   a. Discuss this problem to determine whether it needs to be studied.
   b. If the solution is known, conclude the discussion. If the solution is not known, write the problem on the board for study.
5. Repeat steps 3 and 4 until all problems anticipated have been identified.
6. Follow the normal steps in solving problems until all problems have been studied.

This approach does reduce the list of problems to those requiring further study, but it has several major weaknesses:
1. It does not give the students an overall look at the entire problem area prior to the discussion of the individual problems.
2. It involves, or should involve, a second discussion of the problems listed in order to provide adequate guidance for teaching / learning activities.
3. There is a strong tendency to accept correct solutions to problems without further study even though the students do not adequately understand the reasons behind the solution.
4. It may discourage some of the less able students from participating because the problems they raise are never "important enough to be put on the board for further study anyway."

5. Any summary at the conclusion of the study of a unit or problem area would probably be limited to the problems studied, thus excluding the problems for which solutions were arrived at through discussion only.

## VARIATION NUMBER FOUR

This variation is an extension of the problem-solving approach presented in this publication. It involves using a problem-solving, teacher-student planning approach in developing the course of study for the entire year, as well as in studying the various problem areas.

This variation is discussed briefly in Chapter 7. The steps in this variation are as follows:
1. Lead the students in discussion to develop a list of the enterprises or activities which should be included in the course for the year.
2. Lead the students to determine the units and problem areas, within each enterprise or activity, which should be taken up during the year.
3. Determine, with the students, at what time of the year the units and problem areas should be studied.
4. Follow the steps in problem solving, as presented in this publication, for each of the units and problem areas identified for study.

While this approach does provide for a very high degree of student involvement, it is not recommended for the following reasons:
1. Planning the course of study requires professional training and expertise.
2. Students do not have the breadth of knowledge needed to contribute fully to the identification of all course content.
3. Students are in class to develop abilities other than those required for teaching or building a course of study.
4. Too much class time is needed for developing a course of study with high school students.

## VARIATION NUMBER FIVE

The last variation to be mentioned here is the one which is presented in Chapter 12 as the approach to be used with young farmers and adults. For this reason, no description of it is given here.

The basic principles and procedures of this variation are the same as for the problem-solving approach already presented in detail. It is, in fact, an adaptation of the approach presented—an adaptation which makes it suitable for older, more mature individuals who are capable of considering more complex problems with less reliance on the guidance of the instructor.

This variation has been used successfully with both high school seniors, and in some instances, high school juniors. If these advanced high school students show evidence of an ability to assume the additional responsibilities involved, there is probably no reason why it should not be used. It should not be used with students who are not familiar with a problem-solving approach unless there is a very careful orientation plus close supervision and guidance during the learning stages.

There are many other variations in the use of problem-solving techniques in teaching. Those described here were selected because they seemed to be representative of the approaches used by many teachers. An attempt has been made to combine the best features of the many variations into a simple, straightforward approach which is flexible enough internally to give every teacher the necessary freedom for exercising individual initiative and yet provide a strong, basic framework to serve as a guide for effective teaching.

# CHAPTER 10

# Using Problem Solving to Control Discipline

ONE OF THE MOST INTERESTING and controversial subjects discussed by school teachers and laymen alike is that of how to obtain and maintain discipline in the school. The teacher is expected to maintain good discipline and, if discipline problems do arise, to handle them in the manner least disruptive to the teaching / learning process. Actually, preventing and controlling discipline within the classroom or laboratory is not unlike the problem-solving approach to teaching presented earlier in this book.

Nearly everyone will agree that discipline problems can arise from a variety of causes—from a home situation; from within the individual because of a health problem, a personal problem, or a personal conflict; from activities outside of the school and home; or from something within the school environment. Also, nearly everyone will agree that very little teaching and learning will result without discipline in the classroom. However, the points of view on how to handle discipline problems are as varied as the kinds of behavior which lead to a discussion of them. The place to start in a discussion on discipline is to understand the situation or factors that influence discipline from the standpoint of both teachers and students.

## MEANING OF DISCIPLINE

In order to discuss discipline in the school profitably, some understanding as to just what is meant must be formed. The *Dictionary of Education*, edited by Carter V. Good, provides the following definition of school discipline:

... the characteristic degree and kind of orderliness in a

given school or the means by which that order is obtained; the maintenance of conditions conducive to the efficient achievement of the school's functions.

The foregoing definition should serve all teaching situations. It implies, at least, that there is a kind of orderliness in most schools which permits efficient functioning of those schools, and that there are means by which such order can be achieved and maintained. With this definition to serve as a guide, the place of discipline in the curriculum and how to obtain and maintain discipline in the school can be discussed meaningfully.

**Background of the Problem**

In considering the place of discipline in the curriculum, a quick look at one aspect of the kind of world in which we live is needed—that dealing with the written and unwritten laws which govern all conduct. There are very few things that individuals do which are not controlled in some way. Some controls are in the form of laws, such as traffic laws and property laws. Some controls are in the form of social mores, illustrated by the social pressures which control language, personal habits, and actions at social functions. People may not like to obey these laws, but, for the most part, they do obey them. People who refuse to obey are considered lacking in self-discipline. Thus, society decrees that individuals should be capable of self-discipline; that the schools are, at least in part, responsible for helping individuals develop the ability to discipline themselves so that they can get along in society; and that discipline, therefore, should be included in the curriculum of the school. How this shall be done is left largely to the teaching profession and is considered in the following paragraphs.

## TO OBTAIN GOOD DISCIPLINE

In general, the teaching profession has approached discipline problems from the standpoint of both prevention and control. Preventive measures have stressed those kinds of teaching procedures which help to create a school environment that is largely free of the causes of student misbehavior. Examples of such teaching procedures are:

1. Teaching in such a way that both the teacher and the stu-

dents always know what work is to be done and the reasons for doing the work. Variety in teaching techniques is also very important.
2. Following a definite routine for daily matters such as roll-taking and daily announcements and beginning and ending classes promptly.
3. Making it a practice at the beginning of each class to check on the light and temperature conditions of the classroom or laboratory and on the classroom seating arrangements.
4. Including specific units of instruction (as a part of the course of study) on topics such as the kind of behavior acceptable in school, how to get along with others, and how to get the most out of school. Too often teachers take it for granted that *someone else* has taught their students how to behave. A problem area on appropriate behavior could be

DO NOT TAKE FOR GRANTED THAT STUDENTS KNOW HOW TO BEHAVE

taught and the leading question could be, "Why is it important to act properly in school?" The teacher could continue by identifying appropriate behavior.
5. Establishing rules early for orderly conduct of the class. The making of the rules for the class can be shared with the students. Above all, the teacher should be certain that the rules are reasonable.
6. Consistently and impartially treating all students fairly. Teachers can ill afford to "lose their tempers" when dealing with trying situations.
7. Learning as much about each student as possible from all available sources of information. This should include a knowledge of the home situation.
8. Constantly challenging students by giving them as much responsibility as they are ready to accept.
9. Praising and rewarding students to promote good conduct. This will also help develop good morale in the group.
10. Being alert to everything going on in the classroom. Nothing is more challenging to the mischievous student than a teacher who appears unable to see more than one thing at a time.

There are undoubtedly many more examples of good teaching procedures which would illustrate the point. Anything done which helps provide students with an educational program fitting their individual interests and abilities will tend to prevent misbehavior.

## TO CORRECT POOR DISCIPLINE

Since preventive education and procedures are not always successful, additional plans in the way of control measures are needed. Use of the more severe control measures should be preceded by a careful study of the entire situation, including a thorough case study of the student being disciplined. Some of the more familiar control measures are:

1. *Ignoring minor classroom misbehavior.*—This control measure is used to a greater or lesser extent by all good teachers. Many teachers, however, hold it in disrepute because of the difficulty of knowing just which types of misbehavior may be safely ignored and which types must not be ignored.

2. *Using simple classroom control measures early.*—Teachers should be alert to everything going on in the room and by using a warning look, a shake of the head, or an immediate question, call attention to their disapproval. Many teachers recommend this kind of control rather than ignoring misbehavior as the first step. It is, without doubt, one of the best ways of keeping the classroom situation under control.

3. *Taking away privileges.*—This is slightly more severe than the first two control measures but well within the understanding of all students. It is also one of the most common control measures used outside of school.

4. *Removing the misbehaving student from the situation.*—The student who misbehaves may be removed from the classroom or laboratory or moved to a new location. Students involved in either case may lose a certain amount of control over their own movements.

5. *Sending the misbehaving student to the principal.*—There should be a definite understanding between teacher and principal regarding the kinds of misbehavior which should be referred to the principal. There should be as few such referrals as possible; too many referrals weaken the control of both teacher and principal and are definite signs of weakness on the part of the teacher.

6. *Suspending and expelling.*—These are severe forms of loss of privilege, denying a student the privilege of attending class or school. Such action cannot be taken without consultation with the administrator.

## MEASURES TO AVOID

Control measures to be avoided include public reprimand, enforced apology, group punishment, threat and humiliation, corporal punishment, and punishment involving activities which students are supposed to do as a part of their school work. School work type punishments include outlining pages of books, writing themes, and other "extra" schoolwork assignments. Such "punishments" tend to create strong dislikes for school and school work.

## DECIDING ON THE CONTROLS

Perhaps more important than the kind of control measures to use is the means by which a particular control measure is determined. Special conferences are particularly helpful for this purpose and should be brought into the picture when simple classroom control is not effective. The first few conferences should involve only the student and the teacher. Applying the problem-solving approach, the teacher should ask four basic questions which should serve as a focal point from which conversations about the discipline situation evolve. These questions are:
1. What did you do?
2. Why did you do it?
3. How did it help you?
4. What can be done to prevent it in the future?

These questions should be asked in numerical order, and question 2 should not be discussed until the student answers question 1. The following clarification of the purpose of each of these questions will assist the teacher in conducting an effective student / teacher conference:

*What did you do?*—Formulating a response to this question will point out to the student the inappropriateness of the behavior being considered. The teacher should not identify the inappropriate behavior. The time for the conference should be arranged immediately after the student misbehavior is noticed. For example, if a student has on several previous occasions disturbed a classmate, and this behavior is noticed again, the teacher should immediately arrange to see the student after class or at some other time during the day. The student should be aware of the reason the conference was set and should be able to relate to the question, "What did you do?" If the student answers the question with, "I don't know," the teacher should ignore the student and go about the duties of the office. This gives the impression that the teacher has lots of time to wait for a response, and the student will begin to think about the activities being missed in other parts of the school. The teacher will spend less time waiting on student answers if the conference is arranged to take place immediately before the student's lunch, at the end of the school day, or during athletic practice. Eventually, when the student does answer, it is an admission of guilt.

# USING PROBLEM SOLVING TO CONTROL DISCIPLINE  183

STUDENT-TEACHER CONFERENCES CAN BE EFFECTIVE
FOR CONTROLLING DISCIPLINE PROBLEMS

*Why did you do it?*—Since the event was an action of misbehavior, the student can find no logical or justifiable reason for the action. This serves to point out to the student that the action was not acceptable behavior. Thus, a response of, "There was no good reason for doing it," is an acceptable answer to the question.

*How did it help you?*—The response to this question would be similar to the answer for the preceding question. The student receives no benefit or value for the inappropriate behavior. Again, this question serves to point out to the student that the behavior was unacceptable. It may be useful to have the student explain the effect of the behavior on other students, on other school personnel, and on parents or guardians.

*What can be done to prevent it in the future?*—This question serves to encourage the student to set the standard for future conduct. It should be the student's suggestion, not the teacher's. However, the teacher should not accept a suggestion that does not address the problem at hand. An effective conference is one in which the student promises not to repeat the behavior and ends the conference by shaking hands with the teacher to show good faith. If the student is unable to suggest appropriate behavior for the future, the teacher may offer several suggestions and let the student choose the most appropriate according to the student's viewpoint. If a student-teacher conference fails to produce lasting effects, then a conference involving the parents may be necessary. (In such cases, the teacher should advise the administrator of the plans.) If this also fails, then control of the situation should shift to the administrator. Conferences such as those indicated will be largely ineffective unless the student and parents do the talking. The teacher's role is to guide the thinking of the student and / or parent through skillful questioning so that all persons involved will develop a clear understanding of the problem situation and accept it.

It is best if the suggestions for correcting the situation come from the student. At least this indicates that the student has partially accepted responsibility for creating the necessary adjustments. If the student's suggestions are not successful, a follow-up conference will provide sufficient opportunity for a re-examination of the situation with corrective suggestions coming from the teacher. This kind of conference procedure treats misbehavior as it should be treated—as an educational problem requiring a solution which, to be effective, must come from the person with the problem. It always should be remembered that it is the student, not the teacher, who has the problem.

## DECIDING WHEN PHYSICAL FORCE IS JUSTIFIED

Of course there are always some exceptional cases for which physical force may be needed. These cases involve situations in which the teacher is being attacked, other students are in danger of physical harm, or a student is in danger of self-inflicted physical harm. For any other cases, the use of physical force is apt to be the beginning of the end of effective classroom management for the teacher involved.

## SUMMARY

In the final analysis, then, teachers should look for and seize upon disciplinary situations as opportunities to develop in students the ability to discipline themselves. The development of the ability of self-discipline is an educational problem that should be planned for as carefully as teaching about subject matter is planned. There are a great variety of preventive procedures and control measures available for use by teachers. However, the all-important consideration must always be the degree to which the discipline situation is used to further the development of good citizens for a democratic society.

◆ CHAPTER ◆

11

# Practices for Conducting an
# ♦ Effective Agricultural ◀
# Laboratory Program

THE LABORATORY SEGMENT of an educational program should contribute significantly to the desired learning outcomes expected of the student. However, as with classroom teaching, teachers must also plan for laboratory instruction and supervise laboratory activities if this resource is to serve its planned purpose.

Laboratories are not designed to be a release valve for the teacher who had made no plans for the upcoming class. Students do not like to wander around the laboratory with nothing to do but wait for the period to end. Furthermore, students who do not have planned activities will find something to do, usually to the detriment of their learning opportunities or to the detriment of other class members. A lack of meaningful activity can also develop into discipline problems for the teacher.

The key to effective laboratory instruction rests with planning, instructing, supervising, and evaluating. Teachers need to plan and to coordinate laboratory instruction with the total agricultural education program. This coordination will help in the development of an instructional program well suited to the needs of the students. Furthermore, laboratory activities will then become a valuable asset to the program and to the students and teacher.

Once instructional activities in the laboratory are under way, supervision is essential to assure that the learning activities will be effective and efficient. Students need guidance while working with equipment and tools in the laboratory, and this provides the teacher with the opportunity to give individual assistance during the period.

Evaluation of instruction in the laboratory is also essential. Evaluation of laboratory instruction should be an integral part of the total evaluation made within a program. Strengths and weaknesses of the laboratory can be identified, and these findings can serve as a basis for making decisions for improvement of future laboratory learning experiences.

Throughout this chapter, certain practices are presented which can assist the teacher in providing effective laboratory instruction. The importance of using these practices is highlighted and examples for their application are provided for the teacher.

*Practice Number 1.*—Have students report to the classroom at the beginning of the laboratory period.

A short period in the classroom can be used by the teacher to check the role, make announcements, and determine if all class members have assignments for that particular day. This meeting can also provide an opportunity for a review of safety practices including the correct use of tools which the students did not practice in a previous laboratory period. In order to allow maximum use of the laboratory period, this session should be as brief as possible. Taking care of these matters at the beginning of the laboratory period is more effective than waiting until the end of the period. By waiting until the end of the period, the teacher may forget to make important announcements or to discuss pertinent business with the students, lack sufficient time to cover the topics, or find the teaching efforts ineffective because of the passage of time before the next laboratory class.

*Practice Number 2.*—Make certain every student has work to do.

Students are less inclined to create discipline problems if they are actively involved in learning activities. Students should not be permitted in the laboratory until they have planned, purposeful activity for that day. In some situations, projects will exist which require that two or more students must work together. The teacher should use careful judgment in determining which students can successfully work together. The teacher may also need to prevent students from working in groups all the time because some students will permit others to do all the work. It is recommended that every student be required to complete at least one appropriate individual activity during each laboratory session.

These individual activities will help the students to improve their skills.

The laboratory period is an important segment of the agricultural education program and should be used to instruct students in the correct operation and use of tools. There will be days when some students will not have a project. The teacher should keep a list of things which need to be completed around the laboratory such as drill bits to be sharpened, tool storage carts to be built, plants to be watered, new safety push sticks to be cut, certain areas of the laboratory to be reorganized, plans and bills of materials for future projects to be prepared, areas of the laboratory to be painted, and plants to be repotted. If any student reports to class without a project, the student, assisted by the teacher, can select the project to be undertaken; this project can serve as part of that student's laboratory learning experiences for that day.

*Practice Number 3.*—Wear appropriate clothing.

Teachers, as well as students, should wear coveralls or suitable protective clothing. Dress suits and some school clothes do not lend themselves to working effectively during laboratory periods. In addition to this, the repair or replacement of damaged clothes

STUDENTS ARE MORE APT TO PARTICIPATE IF THEY ARE DRESSED APPROPRIATELY

can be costly. Clothes used in the laboratory should not be ragged, since torn clothing invites accidents around equipment and gives inadequate protection to the wearer. Once clothing becomes soiled and greasy, it should be washed to prevent undesirable odors or the possibility of fire. Leather gloves, appropriate helmets, and goggles should be available for equipment which may present a danger to the operators or to others standing nearby.

*Practice Number 4.*—Require the use of goggles.

Goggles protect the eyes from metal and wood chips, sparks, splashing liquids, or other harmful materials and are required by state laws. The administrator and teacher should exercise care in selecting goggles. They should be comfortable to wear, should not steam up, and should provide adequate protection. Signs placed at the door in the laboratory, in conspicuous places around the laboratory, and on equipment in the laboratory will serve to remind the teacher, visitors, and students that goggles are required by *state laws*.

*Practice Number 5.*—Require the use of safety push sticks.

Each piece of equipment requiring a push stick for safe operation should have a push stick attached to that equipment for easy access. A student will readily use a push stick when one is available but will not use the push stick if time must be spent in looking for the stick or in walking across the laboratory to retrieve one. To help avoid misplacement of these safety devices, the name of the equipment should be painted on each stick.

*Practice Number 6.*—Provide a suitable working environment.

Students will not work at their best if the laboratory is either too cold or too hot. The temperature should be maintained at a level at which it is comfortable to work in coveralls or work clothes. If adequate fume removal systems are not installed, the overhead door and windows should be opened to provide good ventilation.

*Practice Number 7.*—Keep horseplay and noise to a minimum.

The laboratory instructional program is an organized part of the curriculum in the agricultural education program and should be conducted as such. Learning will be seriously affected if the teacher permits students to run free. In addition to creating an undesirable learning environment, horseplay creates safety hazards. A simple joke can turn into a serious accident. Students

FREE TIME IN A LABORATORY CAN LEAD TO HORSEPLAY

must learn that the laboratory provides them the opportunity to develop skills and not the opportunity to indulge in a social hour. Normally, most laboratories are noisy, but students should not yell from one end of the laboratory to the other or engage in excessive conversation unrelated to their projects. Sound levels should be compatible with the nature of the activities being performed.

*Practice Number 8.*—Provide instruction in the use of fire extinguishers.

In case of fires, students may be able to control the fire before the teacher can get to that location. For this reason, students should receive instruction for the safe operation of fire extinguishers. This instruction should be repeated each year for each class. Students should also be familiar with the evacuation procedures of the classroom and laboratory and with what action should be taken if a fire does occur.

*Practice Number 9.*—Supervise the laboratory at all times.

The teacher is responsible when students are in the laboratory and leaving the laboratory unsupervised may lead to problem

LEAVING THE LABORATORY UNSUPERVISED CAN LEAD
TO SERIOUS PROBLEMS

situations. Supervising the laboratory is a full time job and requires the teacher's full attention. Personal telephone calls, individual conferences, or other tasks should be planned or carried out during the teacher's planning time or after school. In the event the teacher *must* leave, all work should stop until the teacher returns or another supervisor arrives.

*Practice Number 10.*—Avoid talking with students while they are operating laboratory equipment.

Drawing the student's attention away from the performance of work in the laboratory could cause a personal injury or damage to project material. The teacher should wait until the student has finished the task and has turned off the machine before talking, unless the teacher perceives a dangerous situation developing. Students should also follow the same practice when they want to talk with another student.

*Practice Number 11.*—Maintain all tools and equipment in safe working condition.

The teacher should periodically check electrical switches, electric cords, oxyacetylene hoses, guards, lawn mowers, chain saws, and other equipment for damage or unsafe conditions. Dull tools are difficult for students to use properly and certainly will affect the quality of their work. Tools or equipment which need to

# AN EFFECTIVE AGRICULTURAL LABORATORY PROGRAM

be repaired or sharpened should be removed from student use until needed maintenance has been performed.

*Practice Number 12.*—Avoid sharing the laboratory with another class during the same period.

Two classes should never share a laboratory during the same period. In multiple teacher departments, the program should be coordinated so that while one teacher is in the laboratory, the other(s) are in the classroom(s). Having two classes in the laboratory at the same time, whether the teachers are agricultural teachers or industrial arts teachers, will create problems. Students will lose project materials, individual student supervision will decrease, the number of tools or equipment available for projects to be completed will be insufficient, the effectiveness of laboratory demonstrations will decrease, and general confusion will increase. Student morale will also decrease when students need to share *their* laboratory period with another class.

*Practice Number 13.*—Remove from the laboratory old equipment, desks, chairs, or other objects not being used.

A laboratory should not be used as a storage room for the school or as a garage for school equipment. Equipment that is not being used (desks, chairs, etc.) takes up valuable space, space which could be used by the student to build larger projects. Old and obsolete items and equipment should be removed. Many laboratories have sufficient floor space, but the accumulation of old equipment and objects makes the facility inadequate.

*Practice Number 14.*—Avoid allowing student use of the laboratory during the teacher's planning period.

Planning periods are for planning or counseling individual students, not for supervising students working in the laboratory. Students like to get out of another class or study hall and go into the laboratory to work on their projects. But the time for working on laboratory projects is during the regularly scheduled class period. If it is necessary for students to work extra periods in the laboratory, the teacher should never leave them unsupervised. Use of the laboratory for a student to gain occupational experience should be planned to occur at times the teacher has scheduled for occupational experience supervision.

*Practice Number 15.*—Use demonstrations regularly.

The laboratory is an ideal learning situation for the students.

Students are using tools everyday and they can experience the value of using tools correctly. Demonstrations can be conducted on an individual basis or with the entire class. If a student is observed using a tool or piece of equipment incorrectly, the teacher should demonstrate to the student the correct use of that equipment. If a class is to be in the laboratory for a period of time, the teacher should plan to give demonstrations on a regularly scheduled basis and prior to student need for the skills / operations to be demonstrated. There are hundreds of tools in the laboratory, and there is a correct method to use each of them. Demonstrations should be given for hand tools as well as for powered equipment. Any student who doesn't have a project for that day can practice using the tool for which the teacher has just demonstrated the correct use.

*Practice Number 16.*—Provide storage for all tools and equipment.

If tools are lost or damaged due to improper storage facilities, the teacher, not the students, is to blame. Every tool should have a designated place for storage. Students are more apt to replace tools which have an established storage place than tools which do not. Wall panels, tool carts, or tool rooms can be used separately

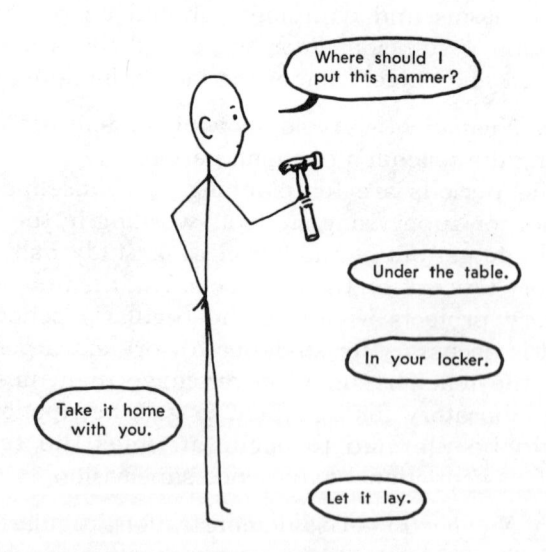

IMPROPER STORAGE ARRANGEMENTS WILL RESULT IN LOST TOOLS

# AN EFFECTIVE AGRICULTURAL LABORATORY PROGRAM

or in combinations for tool storage. Once every tool has a storage place, especially if identifying silhouettes have been painted, the teacher can tell at a glance if all tools have been returned. Tools should be locked up after each laboratory period to minimize the chance of loss.

*Practice Number 17.*—Assign space for storage of projects under construction.

Project storage space should be provided if more than one class uses the laboratory. The possibility of material being damaged or stolen will be reduced if an assigned area is set aside for uncompleted projects. Students should be familiarized with these assigned areas. Space for storage may exist under benches, along the walls, overhead, or in outside areas.

*Practice Number 18.*—Provide space for storing goggles and work clothes.

If lockers are available, one should be assigned to each student. Students should return their work clothes to their lockers after each laboratory period. If no system for storage is arranged, students will lay their goggles and clothes on bench tops, in the corner, or on the floor. This situation clutters the laboratory and creates fire and safety hazards. If lockers are not available, clothes hooks should be provided for clothes. One of the arguments students use to avoid bringing clothes is that they are always stolen. A good storage arrangement would help to solve this problem.

*Practice Number 19.*—Avoid serving as general mechanic for the repair of school equipment.

A teacher's time is too valuable to be used fixing school equipment. Schools have a certain amount of the budget set aside for repairing equipment. Once a teacher starts to repair equipment, each year will bring more and more repair projects.

*Practice Number 20.*—Maintain a clean laboratory.

A cluttered and dirty laboratory creates a bad image of the program in the eyes of the administrators, the parents, and the general public. Administrators will take pride in the program if the laboratory is reasonably clean, and they will tend to support the program to a greater extent. Students will also react with more respect to the equipment and to each other in a clean laboratory.

198    TEACHING AGRICULTURE THROUGH PROBLEM SOLVING

*Practice Number 21.*—Establish a regular cleaning schedule.

Teachers should not be expected to clean up after students. An established cleaning schedule will save the teacher from the task of policing the laboratory after each class, sweeping up sawdust or dirt, hosing down a greenhouse, or returning tools to the storage area. Students do not mind cleaning up the laboratory, but they need some established pattern to follow. One possible cleaning schedule is illustrated in Figure 11-1.

The example in Figure 11-1 is specifically for a class of 14 students. Different cleanup tasks can be increased or decreased in

**FIGURE 11-1**

**Laboratory Cleanup Schedule**

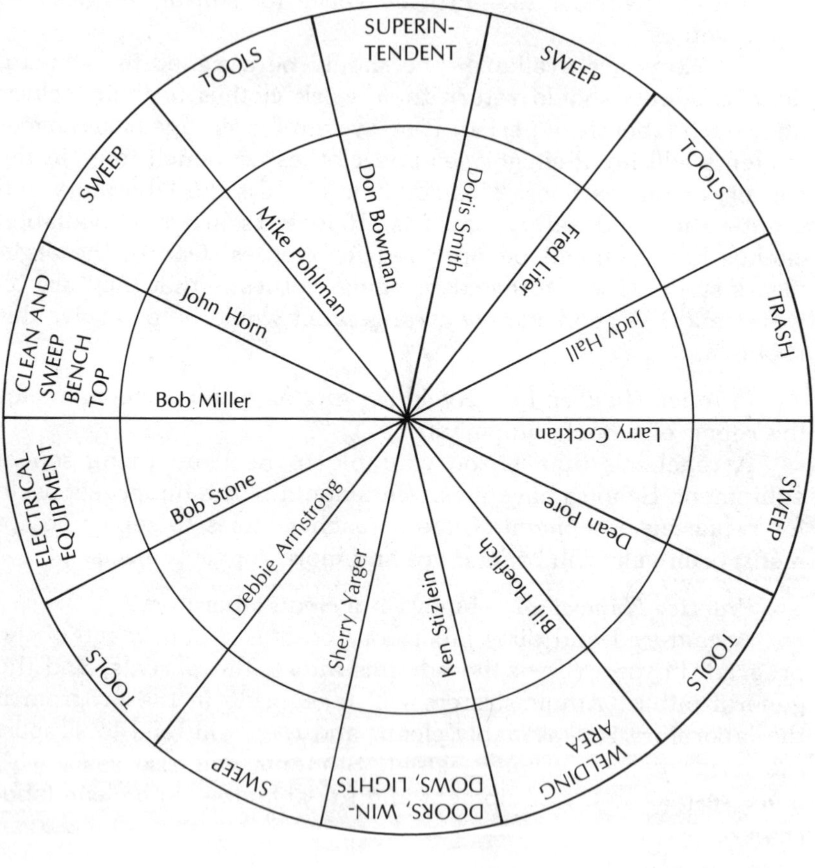

order to involve all students. After each laboratory class, the inner wheel should be moved one space in a clockwise direction. In this way, all students will share in the cleanup and be assigned a particular task. To make this device work effectively, all students should help with the cleanup, even if they have not worked in the laboratory that day. A brief explanation of each cleanup task follows, and it is important that all students understand the responsibility involved with each task.

"Cleanup" responsibility does not mean that those students assigned do all of the cleanup work. All students should clean up their own area and then help others. The cleanup assignment is to place responsibility for making sure the task is actually completed and doing as much as is needed of the actual cleanup work.

*Superintendent.*—The superintendent announces a cleanup time five to eight minutes before the end of the period and is responsible for seeing that each class member assumes a roll in cleaning the laboratory. The superintendent checks to be sure that all projects are properly stored until the next laboratory period and is held responsible if the laboratory is not left in an acceptable condition.

*Sweep.*—The student in charge of this task must see that the entire laboratory is swept and that brooms are returned to the storage area. In an agricultural mechanics laboratory, a sweeping compound should be used once a week to remove fine dust from the floor.

*Tools.*—The assignment for this task involves making certain that all tools are returned to the proper storage.

*Trash.*—Supervision of this task includes seeing to it that all dirt and debris is collected and placed in the proper receptacle.

*Welding area.*—The responsibility for this task includes policing the welding area for unusable metal and having it removed, rearranging hoses and electrical cords, and making sure that all welding and oxyacetylene equipment is properly turned off.

*Doors, windows, lights.*—The student assigned to this task is responsible for making certain that all doors and windows are closed and that all lights in the laboratory are turned off. It is the teacher's responsibility to lock the overhead doors, windows, and the tool storage area after the tools have been returned.

*Electrical equipment.*—This task requires the student in charge to make sure that all electrical equipment is turned off at the end of the laboratory period. This task also includes checking the equipment for unsafe conditions and any other maintenance which needs to be performed on the equipment.

*Clean and sweep bench tops.*—Project activity during the laboratory period will result in sawdust, dirt, soil mixture, scrap lumber, or other scrap material being left on bench tops. Individuals assigned to this task should make certain that benches have been swept before the floor is swept. They should also check for items missed by other students in storing their projects until the next laboratory period. One member should be designated to be responsible for cleaning the sink.

An alternative to the *team responsibility approach* to cleaning is to hold the student using a certain laboratory area and certain tools responsible for cleaning that particular area and for returning the tools to the proper storage area. Students are expected to assist their fellow students after completing their own cleanup. With this approach, the teacher is responsible for checking the final results.

*Practice Number 22.*—Coordinate classroom learning experiences with laboratory projects as much as possible.

The laboratory period should serve as an important learning experience for class members. Incorporating classroom lectures, demonstrations, or discussions on topics related to current laboratory projects makes laboratory work more meaningful. A few examples would be reviewing instruction in arc welding, drawing plans and estimating bills of materials, debudding roses, and identifying plants.

On some days during laboratory periods throughout the school year, some class members will not have projects. Recent classroom topics or demonstrations can serve as a basis for selecting learning experiences for those students who may not have a project on a particular day. For example, previously, a teacher may have demonstrated the proper procedure for sharpening a lawn mower blade. The teacher could have a student who does not have a project practice that procedure on an old blade. A list of little projects such as this should be kept so that potentially wasted time can be turned into a valuable learning experience when the opportunity arises. (See also Practice Number 2.)

# AN EFFECTIVE AGRICULTURAL LABORATORY PROGRAM

*Practice Number 23.*—Inventory tools and supplies.

The teacher should not wait until the start or middle of the school year to perform this task. A complete inventory before school starts will enable the teacher to obtain needed equipment, tools, and supplies. When school reconvenes, the teacher is then ready to conduct the instructional program. The inventory can also serve to check on missing tools and equipment. Once a complete inventory has been developed, the teacher may need to perform this task only once per year, preferably in the spring at the close of school.

*Practice Number 24.*—Check expendable supplies regularly.

Nothing is so frustrating as to plan a demonstration or some project activity and find that the supplies available are inadequate for the activity. Not only does this situation ruin plans for that day, but it curtails the students' learning experiences in the laboratory until the supplies are ordered and received. Standard supplies should include first-aid supplies and fire extinguishers.

*Practice Number 25.*—Remember that laboratory periods are for students and not for teachers.

Teachers should refrain from working on their projects while students are in the laboratory. Laboratory periods are designed for students. Teachers will find their time fully utilized if they supervise the students properly.

*Practice Number 26.*—Coordinate laboratory instruction with the total departmental program and community resources.

A higher quality laboratory instructional program can be provided for all students if the time in the laboratory is scheduled to avoid conflicts with space or equipment between classes.

Avoid laboratory instruction which focuses upon agricultural equipment which is needed during the seasonal peak loads. The time to teach a unit on sprayers is three or four weeks before the spraying season. Students would have a better opportunity to bring sprayers into the laboratory before spraying is underway in the community.

There are also seasonal considerations in determining laboratory instructional topics in the different options. Plants are potted during certain times of the year, and the teacher should relate activities in the laboratory to the operations in local greenhouses.

*Practice Number 27.*—Evaluate laboratory periods regularly.

Before any improvement can be made in the laboratory segment of an educational program, teachers must consider several factors. By using a systematic evaluation scheme, teachers should be able to make rational decisions in adjusting their laboratory activities in order to improve the learning situation. Steps to follow in evaluation are:
1. Define objectives for each laboratory period or for some other appropriate length of laboratory instruction.
2. Designate the data needed to determine if the objectives were fulfilled. Examples of data might include samples of butt welds, quality of workmanship, correct procedures followed, safe working habits practiced, etc.
3. Collect data identified in Step 2 and other relevant information (also secure student input).
4. Study and interpret data collected in terms of previously stated objectives.
5. Identify strengths and weaknesses of the program in relation to the effect these had in accomplishing or not accomplishing the objectives.
6. On the basis of the first five steps, develop new ideas, changes, and innovations which may serve to strengthen the program.
7. Implement these new ideas.
8. Re-evaluate to determine if the laboratory instruction has been improved or has affected other parts of the laboratory work.

These steps are quite detailed, but program evaluation which omits any of these steps may do more harm than good. Furthermore, just determining what has occurred in the laboratory period with no effort to implement needed improvements will also find the teacher falling short of an improved program.

In summary, some teachers spend as much as 50 percent of instructional time during the year in a laboratory setting. This implies that careful planning must be undertaken to assure that time spent in this part of the educational program is an effective and efficient use of both the students' and the teachers' time.

# CHAPTER 12

# Using Problem Solving in Teaching Adults

THERE SEEMS TO BE NO LOGICAL REASON why sound procedures for teaching high school students should not also apply to teaching adults. Teachers must recognize the differences in the maturity and ability of the persons being taught at the various levels and make any necessary adjustments in the basic procedures. Difficulties arise when the teacher fails to recognize these differences and attempts to use similar procedures for all students. A teacher who has experienced making adjustments in the problem-solving approach in teaching all high school levels will be able to adjust the procedure in teaching adults.

Following are a few major factors necessitating some variations or adaptations of these procedures in the teaching of adults.
1. Adults have more control over what they do.
2. Adults are capable of considering broad, complex problems.
3. Adults are capable of more self-direction in their consideration of problems.
4. Adults require less motivation from the teacher. They are far more eager to take advantage of educational programs in vocational agriculture than schools are to provide such opportunities.
5. Adults assume more responsibility for the extent to which conclusions reached in class meetings are put into practice. This does not mean that the help and guidance of the teacher in making this application are not required. On the contrary, the on-the-job instruction, where planning is specific to existing conditions, is probably the most important phase of the adult education program. In the final analysis, the effectiveness of the adult education program in agriculture will be determined by the effectiveness of the on-the-job instruction.

206                TEACHING AGRICULTURE THROUGH PROBLEM SOLVING

PLAN WITH THE CLASS! DON'T FORCE YOUR PLAN
ON THE CLASS JUST BECAUSE YOU HAVE ONE.

6. Adults not only assume more of the responsibility for the direction in which the class moves, but also they assume more of the responsibility for the soundness of the conclusions drawn at the end of the consideration of a particular problem.
7. Adults provide a much richer source of information and knowledge about agriculture within their own group. There is less need for resource persons outside the group itself. When resource persons are brought in from outside the group, they should be used only in a consultant capacity.
8. Courses for adults are usually publicized in terms of specific course titles. This means that there is a homogeneity of interests since the adult farmer will register for the course which will be of the greatest benefit.

## OUTLINE OF THE TEACHING PLAN FOR ADULT INSTRUCTION

The following is a topical outline of the teaching plan for adult instruction. The differences noted between the plan for teaching high school students and the plan for teaching adults will be clarified in the discussion concerning each part of the plan following the outline.

**Part I. Pre-planning Analysis**
  A. Selection of course title
  B. Analysis of the teaching situation
  C. Definition of the teaching / learning objectives

**Part II. The Teaching Plan**
  A. Course title
  B. Interest approach
  C. Anticipated group objectives
  D. Anticipated problems and concerns
  E. Steps in solving the problems
  F. Evaluation and application

**Part III. Teaching Resources**

### Understanding the Plan

The following is a brief discussion of each part of the teaching plan for adult instruction. It must be clearly understood that this presentation is based upon the assumption that systematic instruction consisting of a series of meetings on one major topic with the same person conducting all of the meetings is being provided. No consideration is given to the "shotgun" type of adult class in which a different topic is presented at each meeting with a different person (probably a speaker) conducting each meeting. The "shotgun" adult class is not the kind of adult program for which teachers of agriculture find it difficult to plan.

PART I. PRE-PLANNING ANALYSIS

*Course Title*

The discussion of problem areas for adult classes in agriculture is based on the assumption that each course will be concerned with a single broad area of activity. That means that the unit or problem area and course title are one and the same.

An examination of several lists of titles of courses offered to adults shows the following to be fairly typical:

| | |
|---|---|
| Beef production | Pasture improvement |
| Swine production | Welding |
| Farm machinery repair | Small engine maintenance |
| Soil conservation | Gardening |
| Poultry production | House plants |
| Crop production | Record keeping |
| Corn production | Lawn care |
| Electrification | |

All of these are on an enterprise, or other broad area of activity, basis. They are suitable "unit or problem areas" for adults. The list is not intended to be complete. The courses to be offered depend on the needs of the adults and must be determined by those concerned in the local communities.

One course title, "Farm Management," appears quite frequently on lists of courses offered to adult farmers. The interpretations which have been placed on this title are so many and so varied that no attempt will be made here to describe what is meant. Instead, an effort to describe this kind of course will be made later in the chapter. It is sufficient to say, at this point, that although farm management courses are broader in scope than the other courses listed, the basic teaching procedures and techniques remain the same.

### Analysis of the Teaching Situation

An adequate understanding of the situation is just as important in teaching adults as it is in teaching high school students. The kinds of information required are quite similar as illustrated by the following:
1. Conditions of the farms and businesses of the class members.
2. The community situation—especially in relation to the major problems faced by workers in agriculture and what is being done about these problems.
3. Local agricultural data.
4. Pertinent state and national agricultural data.

Of particular importance is a knowledge of the home farm and on-the-job situations of the class members. The courses should be set up and enrollments taken far enough in advance of the beginning of the course to permit the teacher of agriculture to visit each of those enrolled and to become familiar with each enrollee's situation and problems.

*Definition of the Teaching / Learning Objectives*

If the teacher has obtained the data necessary for an adequate understanding of the situation, there will be no difficulty in determining the objectives for the course. These will, of course, be based on the needs of the members enrolled in the course. These objectives can be stated in broad terms.

The emphasis always should be on changes in the behavior of the class members. Adults could be experts as far as knowing what approved practices should be used, but they will not increase the income from their businesses by as much as one dollar if they do not change their ways of doing things and put these approved practices into operation.

## Part II. The Teaching Plan

*Course Title*

The title for the course is as stated in the pre-planning analysis. After the teacher has analyzed the teaching situation and has defined the major objectives, it is time to consider planning in relation to what is going to happen in the classroom. The items which follow relate to the actual classroom situation.

*Interest Approach*

The first impressions adults have of a class are more important than the first impressions high school students have of their first class. The adults may not come back if it looks as though their time is being wasted; the high school students will be back the next day whether they want to or not, and the teacher will have a chance to make a better impression. The interest approach for an adult group should, therefore, provide the kind of stimulation which will set the pattern for the entire series of meetings. If the teacher has visited all of the class members prior to the first meeting, the task will be much easier.

In planning the interest approach for an adult group, the teacher must keep in mind the many purposes which it should serve. The more important of these purposes are as follows:

1. To arrange an opportunity for the members of the class to become acquainted with one another.
2. To familiarize the members with the various farming or agribusiness operations and jobs that are represented in the class.
3. To provide for some means of "warming up" the members to the point at which they will enter into discussion freely.
4. To acquaint the members with the purpose of the meetings, the manner in which they will be conducted, and the responsibilities of both the teacher and the class.
5. To "point up" the area of agriculture to be considered in this particular series of meetings.
6. To set the stage for the development of plans by the class members for putting decisions arrived at in group discussions into practice. There is no reason why teachers should fear to broach the problem of definite planning for putting these practices into operation. In effect, these plans become the supervised experience programs of the adults enrolled in the course.
7. To set the stage for the establishment of group objectives.
8. To encourage the teacher to become better acquainted with the members of the group.

It is easy to see that the interest approach must serve a much broader function here than with high school students. With the adult class, there will be only one general interest approach; with the high school class there will be many.

*Anticipated Objectives of Those Enrolled*

The teacher is interested in the changes in behavior of the class members. The class members are interested in the effect of the instruction on the efficiency with which they can operate their businesses or perform their jobs. It is in terms of the interests of the class members that the group objectives will be set. The teacher should anticipate this and have data available showing what the average farmer and businessperson are accomplishing and what the best farmers and businesspersons are able to accomplish. The members of the class will then be able to compare their own accomplishments with these data and arrive at some reasonable objectives.

These objectives will usually be stated in terms of either production goals or profits with the exception of some of those of the agricultural mechanics courses which involve only the development of skills.

It is important that some definite objectives be set by the group. Sound objectives give direction and purpose to an activity which might otherwise deteriorate into a social gathering. They emphasize the serious purpose behind the instruction being provided and the fact that there is a need for making some changes. They help individuals to see their own operation in relation to what others think should be accomplished.

*Anticipated Problems and Concerns of the Group*

It is difficult to anticipate the direction in which the thinking of an adult group may go. The adults will have had a great deal of experience in the "school of hard knocks" in addition to a lifelong exposure to many sources of information regarding the various aspects of their businesses. They will be approaching the analysis of a particular area primarily from an economic point of view and will not be concerned with a thorough examination of all of the problems concerned with the production of a particular crop or the performance of a particular task; they will turn only to those specific problems which they feel are their immediate sources of difficulty. What these problems are will depend on a great many factors. World conditions, national and area market conditions, and the farming and business situations in other states will all play a part in determining the particular problems which will be uppermost in their minds. All of this information must be considered in light of the total situation before the teacher can anticipate what will be important to the adult class.

If this were all the teacher had to go on in order to anticipate the problems of an adult class, it might be well to call off the whole program. Such is not the case. The teacher does have two quite reliable bases for anticipating the problems of an adult class at least well enough to make adequate preparation for it. The first of these bases is the analysis of the enterprises into units and problem areas which the teacher made in order to plan the high school course of study. These problem areas, if properly determined, would be very similar in scope to the problems raised by adult workers in the field of agriculture. The problem area for the high school student thus becomes the problem for the adult. There are, of course, exceptions to this manner of stating the problems of the adult, but it will serve as a partial guide for the teacher to follow in planning. The completeness of the analysis means that the teacher should be prepared to proceed in whatever direction the adult group decides.

The second basis the teacher has for anticipating the problems of the adult is the on-the-job visitation prior to the beginning of group meetings. One of the major reasons for making these pre-class visitations is to enable the teacher to get a fairly accurate picture of the problems with which the adults are faced.

*Evaluation and Application*

Once again evaluation and application are found to be inextricably tied together. Certainly, plans must be made for putting new practices into operation; then the teacher must carry the instruction to the job to assist the adults in putting their plans into effect. The teacher can evaluate the effectiveness of the instruction on the basis of the extent to which new practices are put into effect. However, the teacher also must remember that the adult will be evaulating in terms of increased production, increased efficiency, increased income, and contribution to the kind of life success makes possible.

PART III. TEACHING RESOURCES

The need for knowing what materials dealing with the course are available in the vocational agriculture department is self-evident. Good teachers are not expected to know all of the answers, but they can be expected to know where the answers can be found. A tentative list of the various ways in which to provide a little variety in the instruction should be made. If variety is not planned for, it probably will not be found. Of special importance is having available the results of the latest pertinent research.

## TAKING THE PLAN INTO THE CLASSROOM

If the teacher has been able to develop an understanding of the problem-solving approach in teaching as presented in relation to the instruction of high school students, there should be no difficulty in developing adequate plans for adult classes by following the outline for a teaching plan as just described. Instead of providing illustrative plans, therefore, an effort will be made here to show how the classroom procedures would develop from those parts of the plan which relate directly to the meeting between the teacher and the class.

The first of these illustrations will deal with a course in crop production; the second, with indoor plants; the third, with beef production; and the fourth, with farm management.

In considering the descriptions of possible procedures presented here, the teacher must again keep in mind that the descriptions are given only for the purpose of developing an understanding of basic principles for working with this kind of group. It is not expected that the teacher will be able to predict exactly what will take place in any particular situation, but if there is a plan and an understanding of how to use the plan, the teacher will be able to adjust easily to the desires of the group.

## ILLUSTRATION NUMBER ONE—CROP PRODUCTION

**Step I. Interest Approach**

  A. A brief orientation by the teacher regarding the course.
  B. Self-introductions and brief descriptions of their farm crop production programs by the class members. This information can be written in tabular form on the chalkboard with one column left blank for adding production data when available.
  C. Production data for their crops supplied by the class members.

**Step II. Establishing Objectives**

  A. Compare the production data of the group members with local and state averages.
  B. Compare the production data of the group with the production required to "break even" as far as costs and returns are concerned.
  C. Decide what rates of production should be attained. These then become the objectives toward which to work.

### Step III. Determining the Problems and Concerns of the Group
A. Have the group select the crop it wishes to consider first.
B. Refer to the objectives as compared to the actual production figures and then ask the group to indicate the things it feels should be considered in order to be able to reach the objectives.
C. List the "things to be considered" on the board as they are suggested. Possible problems for a course in crop production are:
1. What is the place of this crop in the rotation?
2. How should the crop be fertilized?
3. How can certain diseases and insects be controlled?
4. What are the best planting procedures—time, rate, seedbed preparation?
5. What are the best varieties for the area?
6. How can harvesting procedures increase yield?
7. What storage and marketing procedures will increase the income from the crop?
8. Are there any ways in which labor costs can be reduced?

### Step IV. Solving the Problems
A. Decide on the order in which to take up the list of problems. This may amount to nothing more than a comment to the effect that the problems are already in a kind of order, so the teacher should proceed from top to bottom without further delay.
B. Initiate discussion of the first problem by throwing it out to the group to find out what practices are now being followed and what, if any, new practices are being planned. Other information can be fed into the discussion by members of the group, by the teacher, or by a member of the group designated to check resource materials which the teacher has available. (The teacher simply gives a group member the material with a request to check on the item under question. The discussion continues while the search for information goes on.) Charts and other aids may be used as needed to provide information. If information is required which is not available, the teacher and group together can decide how to get it. Responsibility for obtaining the information can be assigned to a group member or to the teacher, or it may be advisable to take a field trip or call in an outside resource person.
C. Lead the group to arrive at suitable conclusions with regard to the problem.

D. Initiate discussion of the second problem and continue as described until all problems have been considered.
E. Summarize by developing a list of approved practices to be followed and by considering which ones can be put into practice immediately and which ones cannot be put into practice until certain obstacles (or other problems) are removed. This should be followed by on-farm instruction to assist the farmer in putting the practices into effect.
F. Have the group select the next crop to be considered and repeat the above procedures.

The time element has not been injected into the previous discussion (or the succeeding discussions) because this will vary so much from group to group. It may take less than one meeting to accomplish the initial planning with the group, or it may take all of the first group meeting and part of the second. The length of time which will be devoted to each problem will also vary. The teacher will have to use good judgment in keeping the discussion moving along while at the same time allowing adequate time for the development of understandings.

## ILLUSTRATION NUMBER TWO—INDOOR PLANTS

### Step I. Interest Approach
A. A brief orientation by the teacher regarding the course.
B. Self-introductions and brief descriptions of the plants in their houses by the class members. The names of the members with the names of their plants can be listed on the chalkboard for identification.

### Step II. Establishing Objectives
A. Compare the different plants identified to highlight the most popular plants. Also identify unique or rare plants.
B. Add to the list of plants named by the class members. The teacher should have a list of indoor plants, some of which were not identified by any of the class members, but which some of the members might want to grow.
C. Review the plants. The plants selected by the class for further study in the classes that follow can serve as the basis for developing the objectives for the class.

### Step III. Determining the Problems and Concerns of the Group
A. Refer to the plants and objectives selected by the class and

then ask the group to indicate the things it feels should be considered in order to reach the objectives. (Note: the teacher may suggest that it would be better to cover only one plant at a time or that it may be possible to group several of the plants together.)

B. List the "things to be considered" on the board as they are suggested. Possible problems for a course in indoor plants are:
1. What are the recommended environmental conditions?
2. What soil mixture should be used?
3. Why should soil conditioners be used?
4. What are the possible soil conditioners?
5. What are some other plants that can be grown?
6. How can plants be propagated?
7. How much and when should the plants be watered?
8. How much should the plants be fertilized?
9. What may harm the plants?
10. How are insects, diseases, and weeds controlled?
11. What type of container or pot should be used?

**Step IV. Solving the Problems**

A. Decide the order in which to take up the list of problems. This decision may be left up to the group; however, if it is felt that the problems should be addressed in a particular order, the teacher should inform the class and explain the reasons.

B. Initiate the discussion on the first problem selected and have the students share their ideas, the current practices they follow, and / or special problems that concern them. This will give the teacher an opportunity to assess what is and is not known. From this point, several alternatives could be followed by the teacher in helping the class to solve the problem. Perhaps using a film strip or giving a reading assignment would be helpful.

C. After the class has been exposed to the above, lead a discussion to arrive at suitable conclusions to the current problem under study.

D. Initiate discussion on the next problem and continue as described until all problems have been considered.

E. Summarize by developing a list of approved practices for growing indoor plants. Certain basic principles could be identified regardless of which plant is being grown. The teacher may also share with the class resources for further study or resources which could be referred to when problems arise.

## ILLUSTRATION NUMBER THREE—BEEF PRODUCTION

### Step I. Interest Approach

A. A brief orientation by the teacher regarding the course.
B. Self-introductions and brief descriptions of their beef production enterprises by the class members. This information can be written in tabular form on the chalkboard with one column left blank for adding production data when available.
C. Production data for their beef enterprises supplied by the class members.

### Step II. Establishing Objectives

A. Compare the production data of the group members with local and state averages.
B. Decide what rates of production should be attained. These then become objectives of the group.

### Step III. Determining the Problems and Concerns of the Group

A. Refer to the objectives as compared to the actual production figures and then ask the group to indicate the things they feel need to be considered in order for them to reach the objectives.
B. List the "things to be considered" on the board as they are suggested. Possible problems for a course in beef production are:
  1. Is it best to buy feeder steers or maintain a breeding herd?
  2. What kind of feeding system should be used for feeder steers?
  3. How should a breeding herd be fed?
  4. What is the best weight at which to purchase feeder steers?
  5. What grade of market animal will result in the most profit?
  6. What marketing practices should be followed?
  7. How can diseases and parasites of beef be prevented and controlled?
  8. How can the housing and equipment for the beef enterprises be improved?
  9. How do "beef futures" operate? How can they be used in a beef enterprise?

### Step IV. Solving the Problems

A. Decide, through group decision, which problem to consider first.

B. Initiate discussion of the first problem by throwing it out to the group to find out what practices are now being followed and what, if any, new practices are being planned. Other information needed can be fed into the discussion by members of the group, by the teacher, or by a member of the group designated to check resource materials which the teacher has available. Charts and other aids may be used as needed to provide information. If information is required which is not available, the teacher and group together can decide how to get it. Responsibility for obtaining the information can be assigned to a group member or to the teacher, or it may be advisable to take a field trip or call in an outside resource person.
C. Lead the group to arrive at suitable conclusions with regard to the problem.
D. Initiate discussion of the next problem and continue as described until all problems have been considered.
E. Summarize by developing a list of approved practices to be followed and by considering which ones can be put into practice immediately and which ones cannot be put into practice until certain obstacles (other problems) are removed. This should be followed by on-farm instruction to assist the farmer in putting the practices into use.

## ILLUSTRATION NUMBER FOUR—FARM MANAGEMENT

This illustration is included for the sole purpose of discussing one idea of what a farm management course for adults might involve. This is only one idea and should not be taken as the only way in which such a course might develop. What might happen in any specific adult class depends on too many factors which are known only to the local teacher.

The farm management course offering for adult farmers is popular today, especially since microcomputers are available at reasonable costs. It symbolizes the best use of problem-solving techniques in teaching adult farmers. It is also the most difficult course to prepare for from the teacher's point of view. It involves the study and analysis of the entire farming operation, with the instruction growing out of the various needs as shown by this analysis. There will obviously be a tremendous variety of individual problems requiring intensive on-farm instruction. The group instruction will need to be devoted to developing an understanding of the procedures for studying and analyzing the farm business plus instruction in such areas of general interest as are indicated by this analysis.

The following is an attempt to list some of the major questions which must be answered before offering a farm management course for adult farmers.

1. *Does the teacher have a thorough knowledge of the farming of the area?*—It is obvious that the teacher will need to be familiar with the farming of the area. Of even more importance is a familiarity with the home farms of those enrolled in the course. This latter knowledge can, of course, be obtained as the course proceeds.

2. *Is the teacher certain of having the ability to study and analyze a farm business?*—The study of an entire farming program is a very complex problem in itself. Since many teachers of vocational agriculture have not been farm operators, it is very important that some experience be gained in the process of the study and analysis of a farm business through college work in agriculture economics or by making such a study of some local farm using the guidance and advice of persons who have the ability to do the job. Thus, the teacher can learn on the job.

   Of special importance is the ability of the teacher to:
   a. Develop a plan or procedure for making a study and analysis of a farming operation.
   b. Develop a plan for making a record of the information obtained.
   c. Develop a long-time program of farm improvement based on the interpretation of the information obtained. This long-time program of improvement becomes the supervised farming program for the adult farmer.

3. *Can the teacher devote a considerable amount of time to individual instruction on the home farms of the class members?*—The teacher will need to become acquainted with the various farming operations of those enrolled, but this will be just the beginning of home farm visits. The farm management courses will obviously lead to a tremendous variety of individual problems. This means that the group instruction will be devoted largely to bringing about an understanding of the procedures for analyzing the home farm

business and instruction in areas of general interests such as building layout and planning, field and farmstead arrangements, soil conservation, soil testing and soil test interpretation, labor management, planning crop and livestock programs, and other problems common to the group. The many individual problems and, in particular, assistance in making plans for farm improvement and putting these plans into operation will need to be taken care of through individual on-farm instruction. Some of this can be taken care of through individual conferences at the school, but this also requires a great deal of time.

4. *Is the teacher prepared to work with a group of farmers over a long period of time?*—Much of the study and analysis of the home farm business will need to be done over a long period of time. This is especially true when the teacher must start with the development of an adequate record system for the farm. It also will take several years to put many of the changes in the farming program into effect. The teacher would not be completing the teaching task if the farmer were left to put the plan into operation without help.

Because this kind of course varies considerably in nature from the courses thus far described, the illustration which follows is presented in the form of a teaching plan. Additional discussion needed to re-emphasize certain points and clarify others is included in italics and should not be mistaken as being a part of the plan which the teacher should make.

### Course Title: Farm Management

*Farmers who enroll should be given a general idea of the nature of the course prior to their registrations. As suggested earlier, this can be accomplished during farm visits preceding the organization of the course.*

### Situation

*It is impossible to include here an adequate illustration of an actual situation. Therefore, only a list of the kinds of information needed is included.*

1. Soil type and condition.
2. Climatic conditions.
3. The kinds of farming programs best adapted to the soil and climatic conditions.
4. The kinds of farming programs being carried on in the community and on the home farms of those enrolled in the course.
5. A list of the major farm management problems—both recognized and not recognized—facing the farmers. These should also be listed in relation to the farms of those enrolled in the course. *It is not to be presumed that this information will be used to tell the farmers what to do, but rather as a guide to the teacher for planning and working with the farmers.*
6. Local and home farm production data for the major enterprises.

*The information specific to home farm situations can be obtained only through visitation and other personal contact; therefore, the teacher who has been in a community for some time will know much of this information in advance of needing it for a specific class.*

## Teaching / Learning Objectives

1. To develop the ability of the farmers to study and analyze their own farming operations from a farm management point of view.
2. To develop the ability of the farmers to keep adequate records of their farming operations.
3. To develop the ability of the farmers to make a record of their study and an analysis of their home farm operations.
4. To develop the ability of the farmers to make a long-time program for improving their farms, including plans for soil conservation, replanning their cropping system, a soil fertility program, a field and building rearrangement, a livestock program, a marketing program, etc.
5. To develop the ability of the farmers to carry out systematically their long-time programs of farm improvement.
6. To develop the ability of the farmers to solve the many problems in production which will emerge as a result of study and analysis of their farming operation.

7. To develop the ability of the farmers to decide whether or not certain changes were worthwhile from a "personal values" point of view. Increased profits are not the only criterion of success.

These objectives are broad and inclusive. It would be an impossible task to list all of the many specific objectives for this kind of adult farmer course. The teacher needs to think through the broad objectives to be accomplished and let the specific objectives develop as the course progresses.

## Teaching Procedures

### INTEREST APPROACH

1. Orient the class to the nature of the course and the manner in which it will be conducted. This orientation should make use of, and build upon, the orientation provided through the home farm visits which have taken place during the summer.
2. Ask the class members to introduce themselves and give a brief description of their home farming operations.
3. Ask the members of the group to describe any experiences they have had with the study and analysis of farming operations. Bring out the details of the procedures followed.
4. Ask the members of the group to describe the procedures used in the study and analysis of farming operations by neighbors or other persons.
5. Ask the members of the group to tell what was accomplished as a result of the study and analysis of farming operations they described. These should be in terms of increased production, increased efficiency, increased net income, general farm improvement, and personal satisfactions.

It may be that no one in the group will have had the kinds of experiences indicated by the previous questions. This should be ascertained by the teacher in advance so that a different approach can be planned to bring about the desired result. A possible approach would be to arrange for a field trip to a farm where the operator has made a study and analysis of the farming operation and to use the field trip as the basis for the interest approach. A second choice would be to have the farmer discuss personal operation experiences with the group.

## ANTICIPATED GROUP OBJECTIVES

If the group members are anticipating a complete study and analysis of their farming operations—and they should be—they will probably feel inclined to delay specific objectives until their study and analysis shows them where the need for improvement lies. Thus the objectives here will be as broad as the teacher's objectives. The group should set these broad objectives in order to give direction and purpose to their activity. This also will indicate to the teacher how well the group understands what is to be accomplished.

In approaching the group with the question of objectives, the teacher might well include some of the reasoning behind the setting of objectives. For example, the teacher might say, "We've been discussing the nature of this course and what the study and analysis of a farming operation involves. Among the things that should stand out are the completeness of the examination of the farm business and the long-time nature of the program for improvement. We all need to know, in general at least, what we as a group want to accomplish. If we can decide what we want to do, then we will have a pretty good idea of what we will be willing to do in order to get the job done."

The objectives which result may be in terms of financial gain, increase in production, general farm improvement, or personal satisfactions. Once the group has stated objectives which encompass the major purposes of the course, the teacher can proceed with the development of problems and the establishment of procedures for solving these problems.

Some possible objectives are listed below.
1. To develop a long-time program of improvement for the entire farm business.
2. To find out how best to operate the farm with the available supply of labor and capital.
3. To increase the farm net income.
4. To find out how to cut costs without reducing income.
5. To find ways to increase the productive capacity of the farm.
6. To learn how to do a better job of farming.

## Anticipated Problems

*If the interest approach and the setting of objectives have served their purposes well, there should be no difficulty in getting the problems stated. Leading questions will, of course, still be needed. The teacher should try not to stay out of the situation.*

Leading questions: How are we going to go about this business of analyzing our farming operations? What things will we need to know?

We'll need to know:
1. What resources are available—land, labor, capital, and management?
2. What would be the best use of the land?
3. What are the "weak spots" in the business now? Soil? Management? Machinery costs? Labor costs? Capital? Records? Cropping system? Livestock system?
4. What would be the best crop and livestock program for the farm?
5. What records are needed? How to keep them? Should I buy a microcomputer?
6. What should be my total plan for improving the business? Where is the best place to start? How fast should the changes be made?

## Solving the Problems

The basic procedures for solving the problems are the same as for the previous illustrations. They are, however, slightly more complicated since the problems are much broader in their scope and implications. For example, the solution of the problems might begin with a study of the soil and proceed according to the following description. The necessary instruction should precede each step.

1. Testing the soil:
    a. Learning how to take the samples.
    b. Deciding whether to test the soil as part of the group work or to have it tested.
    c. Making soil test maps.
    d. Interpreting the results.
2. Mapping the farm for the following: (Other agencies, such

as the Soil Conservation Service, may be called upon to help.)
   a. Soil type.
   b. Past soil management.
   c. Erosion problems.
   d. Drainage problems.
   e. Field arrangements.
   f. Ditches, fences, etc.
   g. Building layout.
3. Taking the inventory of livestock, equipment, and supplies.
4. Studying farm records to determine major sources of income and to discover income leaks. This might well lead to revising the farm record system.
5. Analyzing all the information about the farm, including management, to locate the weak points or limiting factors.
6. Planning a long-time program of improvement, indicating the order in which the various steps should be taken.
7. Beginning instruction with the problems growing out of putting the long-time program into effect.

As is indicated in these steps, a thorough farm management course will require a great amount of individual work by the adult farmers—the amount dependent upon the adequacy of their records and the extent to which other agencies can be utilized in the study. In any case, a great deal of individual on-farm instruction and counseling will be required. The course will proceed only as fast as the individual farmers are able to proceed.

It is, of course, possible to grow gradually into such a program through courses which are more limited in scope, such as farm records and accounts, making and using soil tests, and soil conservation.

Many of the complications in the instructional program will arise because of the time required for the farm studies to be made. The problems cannot be taken up one at a time and completely solved before moving on to the next one. Soil testing, mapping the farm, and taking the inventory may be going on more or less concurrently. It will also be necessary to plan for some systematic way to record all of the information obtained in such a way that it will be preserved for later use.

As such a course develops, particularly after the first year, the group instruction will begin to take on quite a general nature with a variety of problems being considered in any one year. The instructional program will obtain its continuity and systematic character through the on-farm instruction and the long-time planning.

### Teaching Resources

*Except for the beginning of the course dealing with the study and analysis of the farming program, these will have to be planned for as the course progresses. A great deal of attention should be given to individual needs.*

### Application and Evaluation

*The entire course is based on the development of a longtime improvement program and putting this program into effect. The course will be evaluated in terms of the success of the improvement programs and the personal satisfactions of those involved.*

## A SPECIAL WORD ON YOUNG FARMER PROGRAMS

The same basic procedures will be used for young farmer instruction. The main differences will be in the kinds of problems considered.

If the decision as to whether a farmer belongs in the young farmer program or in the adult farmer program is based upon the degree of establishment, then the young farmer program is left free to consider those problems which deal with getting established. As these young farmers become firmly established, they will gradually drop out of the young farmer courses because these courses will no longer fill their needs. Some production problems will be considered, of course, but not to the extent that they are considered in the adult farmer courses.

The following problems are suggestive of the kinds of problems which might be considered in a young farmer program:
1. Purchasing a farm.
2. Securing satisfactory leases and agreements.
3. Financing the farm business.
4. Planning for the farm insurance programs.
5. Maintaining good landlord-tenant relations.

## USING PROBLEM SOLVING IN TEACHING ADULTS

6. Studying the different ways of getting started in farming.
7. Planning the rotation for the farm.
8. Planning the livestock program for the farm.
9. Setting up an adequate system of farm records and accounts.
10. Studying farm organizations.
11. Providing instruction for problems in the area of farm mechanics.
12. Purchasing farm needs.
13. Marketing farm products.
14. Improving farm living.

As is indicated by the suggested list of problems, some production problems will be considered. However, if the instruction is kept at the level required for getting established in farming, there will be no desire on the part of these young farmers to remain in the program beyond the period needed for getting established.

The main factor in keeping the course of the desired nature will be that of having as the primary agricultural objective "Getting Established in Farming."

The steps in the development of the young farmer program might be as follows:

1. *Interest Approach.*—The interest approach should be based on a discussion of the experiences of the group in getting established in farming.

2. *Setting of Group Objectives.*—The objectives of the group might well be stated in terms of what the members consider establishment in farming to be.

3. *Development of Problems.*—The list of problems would be based on the discussion in the interest approach. If the young farmers are truly concerned with problems in getting established, the list will look something like the suggested list. It should be understood that the planning is being done in terms of a two- or three-year program. The program should be reviewed and modified annually to fit changing conditions.

4. *Solving the Problems.*—The problems should be considered one at a time.

# CHAPTER 13

# Evaluation of Teaching / Learning Activities

IMPROVEMENT OF THE TEACHING / LEARNING PROCESS will never be fruitful unless a critical analysis is made of the teaching currently being practiced. A student teacher is continually observed and evaluated by both the cooperating teacher and the university supervisor. The first-year teacher may be observed by a state supervisor or by a teacher educator as part of a continual improvement in teaching skills. In the latter situation, the observation and subsequent conference sessions are relatively short in terms of the amount of teaching that is carried out by the teacher. The responsibility for improvement of any teaching after an individual accepts a teaching job, therefore, falls back on that individual teacher. Thus, a teacher must become a self-critic, since it is the teacher who is in the classroom or laboratory each day. To aid in this critique process, four separate but related analyses of the teaching / learning process have been identified. These are the analyses of teaching techniques, of teaching, of the teacher, and of laboratory instruction. Each will be discussed in further detail.

## ANALYSIS OF TEACHING TECHNIQUES

Reference was made in Chapter 4 to the analysis or evaluation of the teaching techniques used by a teacher. While it is impossible to identify specific situations in which teaching techniques can be analyzed, there are some unique yet basic criteria that can be used in the evaluation of any teaching technique regardless of the situation in which it is used. These criteria have been summarized in Figure 13-1.

Each of these criteria will be discussed briefly, and implica-

## FIGURE 13-1
### Analysis of Teaching Techniques

| High | Degree of Student Involvement | Low |

| Fast | Speed at Which Information Is Presented | Slow |

| Direct | Role of Teacher | Indirect |

| Many | Resource Materials Needed | Few |

| 50 | Number of Students Involved in Use of the Technique | 0 |

| Competent | Teacher Competency with Teaching Technique | Incompetent |

| 10 | Usage Made of Teaching Technique Within Last 10 Class Periods | 0 |

| Psychomotor | Affective | Cognitive |

Domain Most Closely Related to Teaching Technique

| Discrimination | Manipulation | Recall | Speech | Problem Solving |

Type of Performance Related to Technique

| Seeing | Smelling | Tasting | Feeling | Hearing |

Senses Required of the Student

| 100% | Objectives Achieved | 0% |

tions in critiquing the use of teaching techniques within a specific class situation will be identified.

One way in which teachers could begin to critique their own ability to teach, using the criteria in Figure 13-1, would be to select a class that had been taught recently for review. In reviewing each of the criteria, the teacher should place a check mark at the point on the continuum where it best describes what actually happened within that particular class period in relationship to that specific criterion. For example, in the first criterion on "degree of student involvement," how many of the students were actively involved within the teaching period? 50 percent? 75 percent? 80 percent? or 100 percent? Following this procedure, a check mark could be made on each continuum and after having evaluated the teaching situation, a teacher could begin to assess exactly what happened within that class period in regard to the teaching techniques used.

Another approach would be to have other teachers or individuals observe the class and place a check mark on the continuum where they thought it would best describe what actually happened within that class period. In describing an effective teaching / learning situation, it would be difficult to generalize where a check mark should fall on the continuum. The point to be made is that teachers should be alert to these criteria and if they were applied in their classes and if check marks repeatedly fell within one area of the continuum, then possibly a change is in order. To help illustrate this point, each of the various criteria will be discussed.

**Degree of Student Involvement**

A review of literature will point out that students should be actively involved in the teaching / learning process in order to enhance their learning. Thus it behooves the teacher to involve as many students as possible within each class period. A review of the teaching techniques in Chapter 4 will immediately bring to light that some teaching techniques will naturally involve more students than others. For example, in role playing, all the students in the class may not be actively involved in the role-playing situation and, thus, the check mark may fall toward the right side of the continuum. In a reverse situation, a field trip would involve all students. Thus, if a teacher's teaching ability were to undergo self-analysis and it was found that students were either consistently not involved or were consistently involved at a very low level within the teaching / learning activities, then the teacher should consider those teaching techniques that would tend to involve the students more within the teaching / learning situation.

**Speed at Which Information Is Presented**

A review of the teaching techniques in Chapter 4 will point out that in certain instances the speed at which the technical information is presented to the student will be affected by the type of teaching technique that is employed. In a discussion situation, a teacher may find that information is drawn out of the students very slowly and that in a given period of time, not much information is presented. On the other hand, a demonstration may allow more information to be presented within a short period of time. The point here is that in developing a teaching calendar and subsequently selecting content for teaching, a teacher may be faced with a restricted amount of time in which to cover a certain amount of material. In these instances, teaching techniques ought to be selected that will allow information to be presented to the students at a more rapid pace. This is not to imply that speed should be a major factor in selecting a teaching technique, since speed may reduce the learning efficiency of the student. The teacher must be the ultimate judge as to which teaching technique would be best to present the material in situations in which limited time is a factor.

**Role of Teacher**

The various roles the teacher will play within a certain classroom can range from a direct role at one end of the continuum to an indirect role at the opposite end of the continuum. For example, a demonstration given by a teacher to a group of students would, in fact, be a very direct role that the teacher assumes within a teaching / learning situation. On the other hand, supervised study would be a teaching technique that would allow a more indirect role on the part of the teacher. Whether a teacher selects a teaching technique that tends to be more direct or indirect may depend upon the nature of the subject matter to be covered and what the teacher considers to be the critical needs of the class. As an example to illustrate this point, the teacher may be faced with teaching a unit on leadership development. One way to help develop leadership in students is to give them the opportunity to explore and take the initiative in leading discussion. Thus, if a teacher selects a teaching technique and assumes a more indirect role, students are given the opportunity to help direct the teaching / learning activities within that class period.

**Resource Materials Needed**

The value of this criterion becomes evident when certain teaching techniques are under consideration for use within a class period. Experimentation is an excellent technique in which students are given the freedom to learn by discovery. However, in order to conduct an effective and efficient experiment, certain resource or instructional materials may be needed. If the current financial budgets within the school hamper the purchasing of needed items, the teacher may need to forego the experiment as a teaching technique and select one in which materials are not needed. Many teaching techniques do require certain instructional materials or expendable supplies and thus, if those materials are not available, the teaching technique may need to be altered and the instructional topic be revised or deleted.

**Number of Students Involved in the Use of the Technique**

Early in this chapter, a discussion was given on the degree of student involvement and its relationship to teaching techniques. Some educational games involve all the students in the class, thus promoting an effective way to enhance student learning. However, if the game being considered involves only 5 out of 20 students in the class, the teacher needs to be aware of the limitations of that particular technique and be ready with alternative learning activities for the remaining 15 students. The point to be remembered on the number of students involved is that it may be impossible to involve every single student with every teaching technique that a teacher may need to use throughout a particular instructional unit. However, if this type of situation continues day after day, it follows that certain students in that class will become nonparticipators in classroom activities and eventually become disin-

# 236 TEACHING AGRICULTURE THROUGH PROBLEM SOLVING

STUDENT INVOLVEMENT IS ESSENTIAL TO EFFECTIVE TEACHING

terested in the topic under discussion. Experienced teachers will vary teaching techniques from time to time to permit all students within the class to become actively involved in the learning activities.

## Teacher Competency with Teaching Techniques

One of the basic principles of learning is the principle of practice. This applies equally well to teachers as it does to students. New knowledge will not be gained unless an individual practices that knowledge in relationship to the application of that knowledge to realistic situations. For teachers, this means that in order for them to become competent in the use of various teaching techniques, they must practice and use these techniques for a period of time until they feel comfortable and competent in using each teaching technique. One of the problems that experienced teachers must guard against is falling into the habit of using only selected teaching techniques throughout the instructional year. If

# EVALUATION OF TEACHING/LEARNING ACTIVITIES

much of the teaching were to be analyzed in local schools, the teaching techniques found to be used most commonly would be discussion, questioning, demonstrations, and perhaps testing. These are only four teaching techniques out of the many that would be possible for the teacher to use. The only way that a teacher is going to learn how to use the brainstorming technique in a class is to actually use that technique. The first several times it is used, the teacher may not be satisfied with the results, but it must be remembered that professional competency in using techniques can be perfected through practice.

USING A DIFFERENT TEACHING TECHNIQUE CAN LEAD TO
INCREASED STUDENT INTEREST

## Usage Made of the Teaching Technique Within the Last 10 Class Periods

Perhaps this criterion would be the most revealing of all the criteria applied to a teaching situation. If in analyzing a class, a teacher found that within the last 10 class periods a technique was used 8 times, this would indicate clearly that the teacher needs to explore the use of other teaching techniques in the future. It must also be remembered that certain teaching techniques tend to require the use of certain senses of the students. In a supervised study, the student is basically required to see, and if supervised study is used day after day, the students may not use their other senses. The basic point in this criterion is that teaching techniques should be varied so that students are required to use different learning senses.

## Domain Most Closely Related to Teaching Technique

A review of Figure 13-2 will point out that certain teaching techniques tend to highlight certain learning domains. While some divisions between domains are not as clear as others, it is evident that a teacher does have an opportunity to select different teaching techniques for the different learning domains.

## Type of Performance Related to Technique

This criterion is closely related to the learning domains discussed in the preceding paragraph. The performance that students are expected to exhibit at the end of a class period will center around the five major areas contained on this continuum. If a teacher is continually expecting students to exhibit performance within only one of these five areas, then the teacher needs to explore further what the student should be learning in the instructional topics. Regardless of instructional topics that a teacher has over a period of time, all five of these particular types of student performances have relevance in an agricultural curriculum. It behooves the teacher to identify these specific types of performances and select the teaching techniques that most nearly provide students the opportunity to develop these performances.

## FIGURE 13-2
## Behavioral Emphasis of Teaching Techniques

| Technique | Affective | Cognitive | Psychomotor |
|---|---|---|---|
| Panel Discussion | x | x | |
| Field Trip | x | x | x |
| Education TV | x | x | x |
| Team Teaching | x | x | x |
| Educational Games | x | x | |
| Debate | x | x | |
| Role Playing | x | x | |
| Discussion | x | x | |
| Resource Persons | x | x | x |
| Questioning | x | x | x |
| Supervised Study | x | x | x |
| Demonstration | x | x | x |
| Brainstorming | | x | |
| Experiments | | x | x |
| Programmed Instruction | | x | |
| Pre-tests | | x | x |
| Video Tape | x | x | x |
| Student Reports | x | x | |
| Lecture | x | x | |
| Note Taking | x | x | |

## Senses Required of the Student

In most cases, panel discussions require the students to hear. Not only do experiments require the students to hear, but also they require them to feel, taste, smell, or see. Field trips involve all five senses in relation to the topic under discussion. The point being made again is that different teaching techniques may require different senses. For more effective motivation of students and continued motivation throughout the class period, the overall teaching strategy selected by a teacher should provide the opportunity to use different techniques that involve the various senses.

## Objectives Achieved

One of the final criteria for evaluating any teaching / learning situation is to ask the question, "Were the objectives established for the class achieved?" Two answers would be possible for a teacher to give. (See Figure 13-3.) If the answer is "No," then the

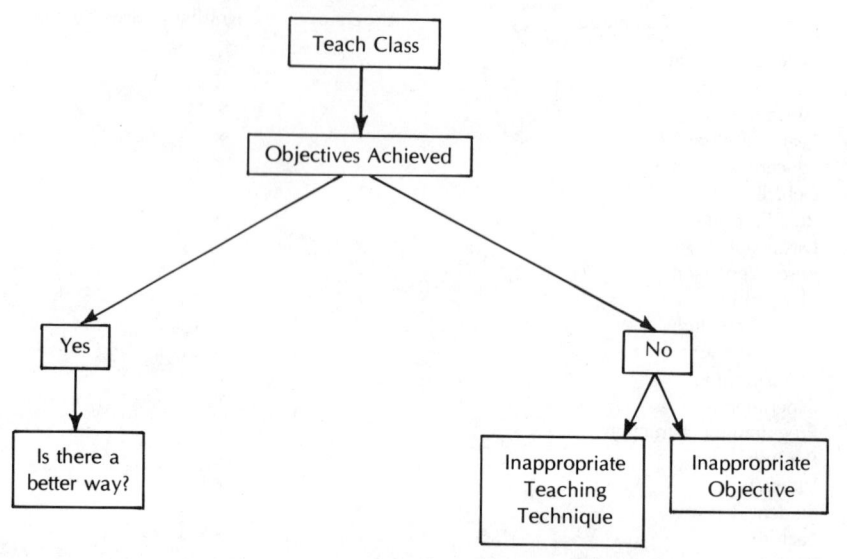

## FIGURE 13-3
### Were the Instructional Objectives Achieved?

```
                    Teach Class
                         |
                         v
                  Objectives Achieved
                   /              \
                 Yes               No
                  |               / \
                  v              v   v
           Is there a     Inappropriate  Inappropriate
           better way?      Teaching       Objective
                           Technique
```

teacher needs to ask if the reason the students did not achieve the objectives was because of the type of teaching technique used. If the answer is "Yes," then the teacher should explore other teaching techniques that would be more effective in reaching the desired objective. The teacher also could have answered "yes" to the question concerning whether the objective was achieved. If this was the case, then the teacher has the responsibility to ask if possibly there is a more efficient way in which to cover the material. The mere achievement of a behavioral objective does not mean that the class was a perfect teaching / learning environment. The teacher must continually seek out better ways in which to approach instructional topics.

If the answer to the achievement of the objectives criterion was "no," then the teacher is faced with finding alternative teaching techniques that will possibly help to achieve the objective. Perhaps a teaching technique was used which was completely inappropriate for the instructional topic. Another factor to consider at this time is that if a class is taught the same subject several times and the objectives are never achieved, then the objectives may be unrealistic for the grade level of the students involved.

Thus, the teacher will need to revise the objectives to bring them more in line with capabilities and interests of the students within that particular class.

## EVALUATION OF TEACHING

The earlier section of this chapter deals with specific evaluation of teaching techniques that a teacher may use within an instructional setting. A more in-depth evaluation of the complete teaching / learning situation must be conducted in order for the teacher to begin to realize what professional improvements must be made. In Figure 13-4, a form is presented that can be used by individual teachers in assessing their own teaching, or it can be used in their observation of others. The form is divided into four major areas: preparation, presentation, teacher, and laboratory work. Provision has been made to evaluate each of the activities on a scale of 1 to 5, with 1 being excellent and 5 being poor.

The preparation section of this evaluation form will tend to assess the pre-planning activities which the teacher should accomplish before entering the classroom. Things considered within this area deal with the lesson plan and quality of the plan prepared. Audiovisual materials deal with the quality of the materials selected or prepared for the topic to be taught. The last major area under preparation deals with objectives, and the question here is, "Does the teacher specifically identify the objectives for a particular class?"

The presentation part of the form deals with that period of time in which the teacher is actually teaching the students. Specific areas deal with the teaching techniques and whether they were used appropriately for the technical information that was to be conveyed. It is important that each class begin with an interest or motivation section, and provision has been made to assess this part of the teaching procedure. The remainder of the activities in the presentation section tend to highlight those things which are consistently used throughout the entire instructional period. At the end of the class period, it is recommended that a summary or review of the materials covered that day be carried out. This can be accomplished through application of the material covered.

Student participation is important to an effective class period. Many times this participation comes through questions asked by the teacher. It would be well if every teacher from time to time

## FIGURE 13-4
## Evaluation of Teaching

Name of unit or problem area taught _____
Date _____ Teacher _____ Course _____ Period _____
Objectives _____

| Activity | Comments | Evaluation* |
|---|---|---|
| **I. Preparation:** | | |
| A. Lesson plan | | |
| B. Audiovisual materials | | |
| C. Objectives (student and teacher) | | |
| **II. Presentation:** | | |
| A. Interest approach | | |
| B. Teaching techniques used properly | | |
| C. Reference materials used properly | | |
| D. Audiovisual materials used properly | | |
| E. Questioning used properly | | |
| F. Chalkboard management | | |
| G. Explanations and illustrations | | |
| H. Class participation | | |
| I. Application of content | | |
| J. Summary and review | | |

*Scale - (1) Excellent; (2) Very Good; (3) Good; (4) Fair; (5) Poor

(Continued)

## FIGURE 13-4 (Continued)
## Evaluation of Teaching

| Activity | Comments | Evaluation* |
|---|---|---|
| **III. Teacher:** | | |
| A. Appearance | | |
| B. Speech (voice, language, etc.) | | |
| C. Enthusiasm, tact, poise | | |
| D. Classroom management (heat, lights, etc.) | | |
| E. Controlling class | | |
| F. Mannerisms | | |
| G. Knowledge of subject | | |
| **IV. Mechanics Laboratory, Greenhouse, Land Laboratory, etc.** | | |
| A. Preparation and organization | | |
| B. Student participation | | |
| C. Quality of work | | |
| D. Demonstrations | | |
| E. Supervision | | |
| F. Safety (equipment and habits) | | |

*Scale - (1) Excellent; (2) Very Good; (3) Good; (4) Fair; (5) Poor

asked somebody to observe the class and record the mode of questioning used. Assume that the "X's" in Figure 13-5 are students and that the subscripts indicate the number of times the teacher directed a question to that student. It is evident that the teacher favored the students to the right. Every teacher should attempt to avoid favoring students at the expense of not calling on others. Many times teachers unconsciously favor one side of the class without realizing it. Class participation also may be influenced by the way the teacher works with the chalkboard. Figure 13-6 illustrates how the vision of the teacher can be limited, depending on how the teacher stands. Teacher A, who works with the left shoulder to the chalkboard, may tend to direct attention more to the left side of the classroom while Teacher B may react in an opposite manner. The major point to keep in mind is that teachers must make a conscious effort to involve all students in the teaching / learning situation and try to overcome any peculiar teaching habits that cause them to omit some students.

The last recommendation regarding student participation concerns seating arrangement. The best seating arrangement is a "U" shape as illustrated in Figures 13-5 and 13-6. This type of seating arrangement is preferred for several reasons.

1. Students will participate in a class discussion more readily when they can see the faces of their classmates rather than the backs of their heads.
2. Discipline problems will be fewer because the teacher can

**FIGURE 13-5**

**Student Participation**

Teacher

| $X_I$ | | | | | | $X_I$ |
|---|---|---|---|---|---|---|
| $X_{II}$ | | | | | | $X$ |
| $X_I$ | | | | | | $X_{II}$ |
| $X_{IIII}$ | | | | | | $X_I$ |
| $X_{II}$ | | | | | | $X$ |
| $X_{II}$ | $X_{III}$ | $X_{II}$ | $X_I$ | $X$ | $X$ | $X_{II}$ |

# FIGURE 13-6
## Student Participation Influenced by Teacher Stance

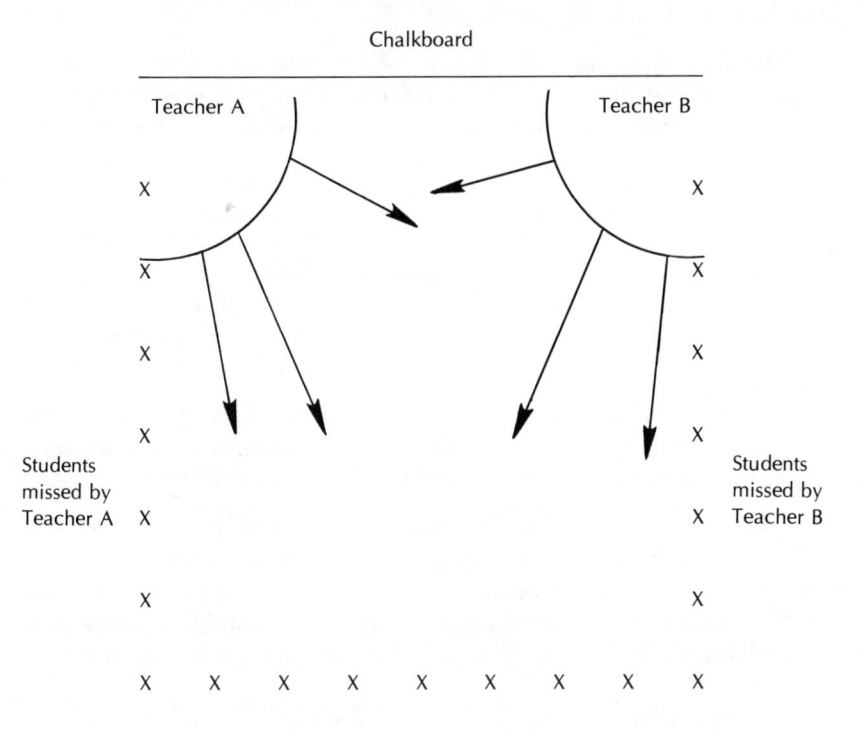

Chalkboard

look directly at the students when there is nothing between them.
3. Audiovisual aids and projectors can be moved in and out of the free area without obstructing the view of the students.
4. The teacher is freer to move about the class to assist individual students.

## ANALYSIS OF THE TEACHER

Again referring to Figure 13-4, there are certain criteria that can be applied to all teachers in assessing those teacher characteristics that relate to quality teaching.

These criteria include:

*Appearance.*—Teachers should present a neat appearance at all times.

*Speech.*—Teachers should use correct grammar and speak in a clear voice to assure that all students can hear and understand the information being presented.

*Enthusiasm, Tact, and Poise.*—Enthusiasm for the topic under discussion and the tact and poise with which they present the topics are all attributes that relate to the confidence that teachers should exhibit in front of the class.

*Classroom Management.*—Classroom management involves those things that deal with the physiological needs of the students; for example, the temperature of the classroom. If the students are too hot or too cold they will not direct their full energies to the topic under discussion. It is the responsibility of the teacher to make every effort to correct this condition.

*Controlling the Class.*—Controlling the class relates to the teacher's ability to identify and prevent discipline problems from occurring. Excellent teaching will tend to discourage misbehavior.

*Mannerisms.*—Distracting mannerisms can, at times, be very disturbing to students and will, if carried to extremes, affect the quality of teaching. Irritating mannerisms such as playing with objects, squeaking the chalk, and uttering a constant barrage of "Ahs" and "Uhs" tend to be very distracting to students and must be avoided.

*Knowledge of Subject.*—Knowledge of the subject is important, and the teacher should not attempt to teach a class until adequate information on the topic has been placed in the lesson plan.

## LABORATORY INSTRUCTION

For the laboratory, a more informal atmosphere usually exists, and the teacher assumes a more indirect role. In this particular situation, a teacher needs to analyze the teaching / learning activities critically to avoid the situation in which a laboratory becomes a wasted period. Other areas to consider are student participation and the quality work that is done within the class period. Demonstrations are frequently used within the laboratory setting and, thus, the technique is highlighted on this particular form. An important part of lab work deals with teacher supervision which involves planned group instruction for the laboratory period as well as individual assistance to those students who need help. Even though this part of the form is rather short in comparison to Section II, it must be remembered that laboratory instruction is important and provides the opportunity for the teacher to convey much technical information to students. It also provides the opportunity for the students to apply what they learned in the classroom and to develop desirable and safe work habits.

# CHAPTER 14

# ♦ The Teacher Is the Key ◄

THIS BOOK BEGINS WITH THE SENTENCE, "The success or failure of teaching often can be traced directly to the effectiveness of planning." The implication of this sentence places the responsibility for this planning where it belongs—upon the teacher. A teacher is hired by the school board to plan and to carry out those educational activities which will lead to an effective program, and this responsibility cannot be passed on to another person, no matter how weak the program is or how weak it becomes. Successful programs of agricultural education begin with effective teaching / learning activities. These serve as a sound foundation upon which other parts of the program can be implemented successfully.

Where strong agricultural education programs exist, good teaching is responsible; where weak programs exist, poor teaching is responsible. Successful programs are a success because they are helping students to solve problems.

The book begins where the teacher must begin—if problem solving is to be used successfully—with the determination of the units and problem areas in the course of study. This step must be preceded, of course, by a thorough study of the agricultural needs of the community in which the school is located.

When the teacher has determined the units and problem areas to be included in the course of study and has identified the potential supervised occupational experience programs, plans can be developed for carrying out effective teaching / learning instructional activities.

A detailed plan is presented in outline form as a guide to the teacher in planning for this instruction. Several illustrative plans are presented. An effort also has been made to show how such a plan would be used in the classroom situation. Many of the "trouble spots" which plague the teacher who attempts to use the plan are described to aid in evaluating what occurs. A minimum plan is

SUCCESSFUL PLANNING LEADS TO THE END OF A GOOD DAY

suggested for those teachers who have developed their abilities to use this approach to such an extent that they feel no longer in need of the detailed plan.

A description of some of the variations in using a problem-solving approach in teaching is provided for those who wish to examine the procedures which differ from those suggested here. It is highly probable that teachers will develop their own variation after using the procedure described here for a while, and this is as it should be.

Following the detailed description of the use of a problem-solving approach in teaching with high school students, an attempt is made to show how the same basic procedures should be used in teaching young farmers and adults.

Several chapters focus upon the relationship of motivation, teaching techniques, and discipline to the problem-solving approach. Each of these respective areas can influence the successful use of problem-solving activities. Teachers must remember that problem solving does not limit the kind or number of teaching techniques that can be used. In fact, problem solving helps to point out to the teacher those techniques and instructional aids which best fit the units and problem areas being taught, whether the instruction is in the classroom or in the laboratory. Finally, the content of this book shows that the use of problem-solving procedures in teaching is strongly supported by the research of many educators who have studied the basic learning processes.

It is impossible to convey through the written word the great amount of faith and enthusiasn which we have for the problem-solving approach in teaching as presented here. Some teachers will have difficulty in using this approach while others will have no difficulty whatever. We can only hope that all teachers will take the time and trouble to analyze their own teaching carefully to determine what the difficulties are, and then seek to correct them.

As was pointed out earlier, teachers must rely upon the "Principle of Practice" to perfect their competence in the use of problem solving. The description of the plan and basic procedures are here. There can be no substitute for the ability, initiative, creativity, and desire to improve on the part of the individual teacher.

# ◆ APPENDIX ◆

# A

# APPENDIX

# A

# Building Courses of Study for Agricultural Education

IN VOCATIONAL EDUCATION IN AGRICULTURE, the teacher has been charged with the major responsibility for developing a course of study to fit the needs of the community. This responsibility grew from the emphasis on production agriculture and the fact that there were major differences in agricultural production even between geographically close communities. This basic philosophy regarding building courses of study is still sound even with the broadening of the program to include preparation for all occupations involving the knowledge and skills of agricultural subjects. It is important that courses of study for vocational agriculture be carefully planned.

## STEPS IN BUILDING A COURSE OF STUDY

1. Study the total agricultural situation of the school district, farm and nonfarm, and determine the agriculture to be taught in terms of enterprises, units, and problem areas.

    Data regarding production, land use, financing, people, occupations, machinery and equipment, the general needs for improvement of agriculture in the area, and the general needs of persons in agriculture are secured from many sources. Some state and national data also have been used to show the relationship of the local situation to the state and to the nation. More recently, information about the nonfarm agricultural businesses in the area surrounding the school district has been added because of the increased emphasis on nonfarm agricultural occupations. Once the agricultural data are available, decisions can be made regarding the programs and enterprises, units, and

problem areas which should be included in the instructional program.
2. Study the school situation. The study of the school situation includes information such as enrollments, facilities and equipment, period length, attitudes toward vocational education in agriculture, past history and program in agriculture, staff, and former courses of study in agriculture.
3. Determine the courses to be offered. Any area of content must be organized some way for instruction. Students wish to know, teachers need to plan, administrators need to schedule classes, and parents want to know. The content in agriculture is divided first into desirable courses and then these courses are reshaped in accordance with the school situation. Much of the decision making in this step is based on steps 1 and 2.
4. Determine the content for each course in terms of units, problem areas, and the time needed for providing instruction for each unit and problem area. Once the courses have been determined, an estimate of the time required for teaching each unit is made, and the units are assigned to the courses to be offered. At this time, if it has not been done before, the orientation and guidance, occupational experience, leadership, and other general units are added to the total list of enterprises and units to be taught. Estimates of time are based on factors such as the abilities to be developed, facilities available, importance of the area to the occupational experience programs of the students, and the agricultural background of the students.
5. Determine when in the year each unit and each problem area are to be taught. Various factors such as seasonality, scheduling of the use of facilities, and the maturity of the students enter into this step.
6. Determine the specific content for each unit and problem area. This is the final step in course building. The content is identified in terms of problems, special events and activities (such as laboratory exercises) only, with the detailed planning for teaching being left as a day-to-day problem.

**PRINCIPLES OR GUIDELINES FOR COURSE BUILDING**

1. The agricultural content of the vocational agriculture series

of courses should be based on a study of the total agricultural industry of the local school district, plus pertinent area, state, and national agricultural data.
2. Local surveys to obtain data should be limited to data not already easily available from sources such as the U.S. Census of Agriculture.
3. The study of a community for building a course of study for an agricultural education program should include a study of the agricultural interests and needs of all of the people in the area served by the school.
4. To the extent possible, data should be recorded when gathered in the form in which they are to be used later.
5. Certain kinds of data should be updated annually, and the course of study should be modified in accordance with the trends revealed.
6. The course of study is built for the community situation and is modified for specific groups of students.
7. The courses to be offered are determined by analysis of the agricultural industry of the area and the agricultural knowledge and skills needed by the clientele for which the program is planned. Adjustments to the school situation are made after the best course structure is determined.
8. The course of study is planned in sufficient detail to serve as a starting point for planning for teaching, ordering materials, ordering equipment, and performing the other tasks needed for conducting an instructional program.
9. The agricultural content is organized into enterprises, units, and problem areas for effective planning and teaching.
10. The principles of learning need to be kept in mind in organizing course content into units and problem areas as well as in assigning content to courses.
11. The objectives for the course of study should be derived from the objectives for the program of vocational education in agriculture.
12. The objectives for the course of study should be stated as educational objectives, not as agricultural production objectives.
13. The objectives for the course of study should be consistent with the objectives for the total program of education for the community.

14. The determination of course content should be based on the objectives for the course of study as well as on the agricultural situation.
15. Units and problem areas for agricultural content should be stated in terms used by agricultural workers to indicate activities performed. To the extent possible, problem areas should reflect seasonal activity and natural divisions in the life cycles of the plants and animals.
16. Major emphasis in the course of study should be given those enterprises which are a part of the occupational experience programs of the students.
17. Areas of study in the course of study should include those needed for the general development and guidance of the student (FFA, occupational experience programs, occupations) as well as the production and management phases of agriculture.
18. Sufficient time should be allocated to each unit and problem area to permit the development of understandings and abilities which will result in the application of what is learned.
19. The course content should involve the student in a study of all aspects of agricultural businesses, both farm and nonfarm.
20. Problem areas of vocational agriculture should be planned so that assignments will come at the time within the farm year that the instruction will be of greatest value to the student.
21. When central emphasis is given to a major enterprise in one year, care should be taken to avoid concentrating all of the units and problem areas for that enterprise in just one or two months of the year.
22. If application of instruction is not possible for high school students, the problem areas may be left for the adult students.
23. The number of problem areas for each enterprise should be varied according to the importance of the enterprise in the community and in the occupational experience programs of the students.
24. Instruction in agricultural mechanics and outdoor laboratories should be arranged so as to provide for effective use of the agricultural facilities throughout the year.

25. The course of study should be flexible to provide for differing needs of students and to make possible adjustments to changes in agriculture and in rural living.
26. The course of study should be evaluated each year in terms of its effectiveness in contributing to the accomplishment of the objectives of the vocational agriculture program.
27. In planning the course of study, consideration should be given to the instructional facilities available and to the administrative organization of the program.
28. With the exception of some units and problem areas in "Orientation and Guidance," assignment of problem areas for formal classroom instruction should be made only once in the vocational agriculture program.
29. The complete course of study should include the following:
    a. Title of course.
    b. List of enterprises, units, and/or problem areas.
    c. Statement of objectives of the course and the relation of these objectives to the overall objectives of the agricultural education program and to the objectives of the total school.
    d. Assignment and time allocations of units and problem areas by courses in agricultural education and by the months in which they are to be studied.
30. Advisory councils may be used to assist in gathering and interpreting the data upon which the course of study is based, and in formulating and interpreting the objectives for the course of study.

# APPENDIX B

## APPENDIX

# ▶ Unit and Problem Area List ◀

THE FOLLOWING LIST of enterprises, units, and problem areas is illustrative only. It is not intended for use in any particular community. An enterprise of major importance in one community will be divided into many units and problem areas. The same enterprise may be of minor importance in another community and may, therefore, be divided into only two or three units and problem areas with a minimum amount of time assigned to each unit. Many enterprises are not included among the illustrations. Enterprises not included can be analyzed into units and problem areas following the patterns indicated in this book.

The list of problem areas also represents much more in terms of content than could be included in the usual one-teacher high school agriculture department program. Certain enterprises and problem areas on the list may be excluded from the program of a particular community.

If other courses are added, such as horticulture courses, advanced agricultural science courses, or suburban agriculture, additional units of content would need to be developed.

## UNITS AND PROBLEM AREAS

**Enterprise: Orientation and Guidance**

Unit A: Orientation to school and vocational agriculture.
Problem areas:
1. Getting acquainted with the high school.
2. Getting acquainted with methods, procedures, and content of vocational agricluture.

Unit B: Relationships with others and with the school.
Problem areas:
1. Getting along with others.
2. Getting the most out of school.

Unit C: Occupational information.
Problem areas:
1. Getting acquainted with types of farming and / or other agricultural businesses.
2. Getting acquainted with agricultural occupations.

Unit D: The world of work.
Problem areas:
1. Getting started in farming.
2. Getting started in a nonfarm agricultural occupation.

## Enterprise: Organizations for Students in Agriculture

Unit A: The FFA.
Problem areas:
1. Getting acquainted with the FFA.
2. Planning programs of work.
3. Participating effectively.
4. Understanding and using parliamentary procedure.

Unit B: Continuing participation in the FFA.
Problem Areas:
1. Evaluating and replanning programs of work.
2. Improving leadership ability.

## Enterprise: Occupational Experience Programs in Agriculture

Unit A: Planning occupational experience programs in agriculture.
Problem areas:
1. Getting acquainted with occupational experience programs.
2. Planning occupational experience programs.
3. Keeping records for occupational experience programs.

Unit B: Evaluating and replanning the occupational experience program.
Problem areas:
1. Summarizing and analyzing occupational experience program records.
2. Replanning the occupational experience program.
3. Managing money earned in occupational experience programs.

## Enterprise: Swine

Unit A: Selecting and buying swine.
Problem areas:
1. Selecting swine by physical appearance.

APPENDIX B

2. Using records in selecting swine.
3. Buying swine.

Unit B: Feeding and caring for swine from breeding to weaning.
Problem areas:
1. Feeding and caring for the sow and gilt from breeding to farrowing.
2. Feeding and caring for the sow and litter at farrowing time.
3. Feeding and caring for the sow and litter from farrowing to weaning.

Unit C: Feeding and caring for swine after weaning.
Problem areas:
1. Feeding growing hogs for market and for herd replacements.
2. Feeding and caring for the herd boar.

Unit D: Housing and equipping the swine enterprise.
Problem areas:
1. Providing housing for swine.
2. Providing equipment for swine.

Unit E: Controlling diseases and parasites of swine.
Problem areas:
1. Preventing and controlling diseases of swine.
2. Controlling parasites of swine.

Unit F: Marketing and improving the swine enterprise.
Problem areas:
1. Marketing swine.
2. Improving the swine breeding program.

Unit G: New developments in swine.
Problem areas:
1. Reviewing and evaluating swine production developments.
2. Reviewing and evaluating swine marketing developments.

**Enterprise: Beef**

Unit A: Selecting and buying beef animals.
Problem areas:
1. Selecting beef animals by physical appearance.
2. Using records in selecting beef.
3. Buying beef animals.

Unit B: Feeding and caring for beef animals.
Problem areas:

1. Feeding and caring for the breeding herd.
2. Feeding and caring for the cow and calf.

Unit C: Feeding and caring for growing and finishing beef animals.
Problem areas:
1. Raising the beef heifer and young bull.
2. Feeding and caring for finishing beef animals.

Unit D: Housing and equipping the beef enterprise.
Problem areas:
1. Providing housing for the beef enterprise.
2. Providing equipment for the beef enterprise.

Unit E: Controlling diseases and parasites of beef.
Problem areas:
1. Preventing and controlling diseases of beef.
2. Controlling parasites of beef.

Unit F: Marketing and improving the beef enterprise.
Problem areas:
1. Marketing the beef enterprise.
2. Improving the beef breeding program.

Unit G: New developments in beef.
Problem areas:
1. Reviewing and evaluating beef production developments.
2. Reviewing and evaluating beef marketing developments.

**Enterprise: Dairy**

Unit A: Selecting and buying dairy animals.
Problem areas:
1. Selecting dairy animals by physical appearance.
2. Using records in selecting dairy animals.
3. Buying dairy animals.

Unit B: Feeding and caring for replacement animals.
Problem areas:
1. Feeding and caring for the cow and calf at calving time.
2. Raising the dairy calf.
3. Raising the dairy heifer.

Unit C: Feeding and caring for the dairy herd.
Problem areas:
1. Feeding and caring for the dairy bull.
2. Feeding and caring for dairy cows during lactation.
3. Feeding and caring for the "dry" cow.

Unit D: Housing and equipping the dairy enterprise.
Problem areas:

# APPENDIX B

1. Providing housing for the dairy enterprise.
2. Providing equipment for the dairy enterprise.

Unit E: Controlling diseases and parasites of dairy animals.
Problem areas:
1. Preventing and controlling diseases of dairy cattle.
2. Controlling parasites of dairy cattle.

Unit F: Marketing the dairy enterprise.
Problem areas:
1. Selecting milking practices for producing quality milk.
2. Marketing milk and cream.
3. Marketing dairy animals.

Unit G: Improving the dairy enterprise.
Problem areas:
1. Testing milk for butterfat and herd testing.
2. Improving the dairy breeding program.

Unit H: New developments in dairying.
Problem areas:
1. Reviewing and evaluating dairy production developments.
2. Reviewing and evaluating dairy marketing developments.

**Enterprise: Sheep**

Unit A: Selecting and buying sheep.
Problem areas:
1. Selecting sheep by physical appearance.
2. Using records to select sheep.
3. Buying sheep.

Unit B: Feeding and caring for sheep.
Problem areas:
1. Feeding and caring for the flock ram.
2. Feeding and caring for the breeding flock.
3. Feeding and caring for the ewe and lamb from birth to weaning.
4. Feeding and caring for growing and fininshing lambs.

Unit C: Housing and equipping the sheep enterprise.
Problem areas:
1. Providing housing for sheep.
2. Providing equipment for sheep.

Unit D: Controlling diseases and parasites of sheep.
Problem areas:
1. Preventing and controlling diseases of sheep.
2. Controlling parasites of sheep.

Unit E: Marketing and improving the sheep enterprise.
Problem areas:
1. Shearing sheep and storing wool.
2. Marketing wool.
3. Marketing mutton and breeding stock.
4. Improving the sheep breeding program.

Unit F: New developments in sheep.
Problem areas:
1. Reviewing and evaluating sheep production developments.
2. Reviewing and evaluating sheep marketing developments.

**Enterprise: Poultry**

Unit A: Selecting and buying poultry.
Problem areas:
1. Selecting and buying chicks.
2. Selecting and buying mature birds.

Unit B: Growing chicks.
Problem areas:
1. Feeding and caring for chicks to 12 weeks of age.
2. Providing housing for chicks.
3. Providing equipment for chicks.

Unit C: Feeding and caring for developing layers.
Problem areas:
1. Feeding the developing layers.
2. Providing housing for the developing layers.
3. Providing equipment for the developing layers.

Unit D: Feeding and caring for the laying flock.
Problem areas:
1. Feeding the laying flock.
2. Providing housing for the laying flock.
3. Providing equipment for the laying flock.
4. Producing eggs for hatcheries.

Unit E: Raising broilers.
Problem areas:
1. Feeding broilers.
2. Providing housing for broilers.
3. Providing equipment for broilers.

Unit F: Controlling diseases and parasites of poultry.
Problem areas:
1. Preventing and controlling diseases of poultry.
2. Controlling parasites of poultry.

## APPENDIX B

Unit G: Marketing and improving the poultry enterprise.
  Problem areas:
  1. Storing and marketing eggs.
  2. Culling the poultry flock.
  3. Marketing poultry for meat.
  4. Improving the poultry breeding program.

Unit H: New developments in poultry.
  Problem areas:
  1. Reviewing and evaluating poultry production developments.
  2. Reviewing and evaluating poultry marketing developments.

**Enterprise: Corn**

Unit A: Planting operations for corn.
  Problem areas:
  1. Selecting and purchasing seed corn.
  2. Applying fertilizer at planting time.
  3. Applying chemicals for weeds and insects at planting time.
  4. Planting corn.

Unit B: Caring for the growing corn crop.
  Problem areas:
  1. Controlling weeds in the growing crop.
  2. Controlling insects and pests in the growing crop.
  3. Preventing and controlling diseases in corn.
  4. Applying supplemental fertilizers.

Unit C: Harvesting, storing, and marketing corn.
  Problem areas:
  1. Harvesting the corn crop.
  2. Drying the corn crop.
  3. Storing the corn crop.
  4. Marketing the corn crop.

Unit D: New developments in corn.
  Problem areas:
  1. Reviewing and evaluating corn production developments.
  2. Reviewing and evaluating corn marketing developments.

**Enterprise: Oats**

Unit A: Growing oats.
  Problem areas:
  1. Selecting and buying seed oats.

2. Planting and fertilizing oats.
3. Caring for the growing oat crop.

Unit B: Harvesting, storing, and marketing oats.
Problem areas:
1. Harvesting and storing oats.
2. Marketing the oat crop.

Unit C: New developments in oats.
Problem areas:
1. Reviewing and evaluating oats production developments.
2. Reviewing and evaluating oats marketing developments.

**Enterprise: Soybeans**

Unit A: Planting soybeans.
Problem areas:
1. Selecting and buying soybean seed.
2. Planting and fertilizing soybeans.

Unit B: Caring for the growing soybean crop.
Problem areas:
1. Controlling weeds in soybeans.
2. Controlling insects in soybeans.
3. Preventing and controlling diseases in soybeans.

Unit C: Harvesting and marketing soybeans.
Problem areas:
1. Harvesting and storing soybeans.
2. Marketing soybeans.

Unit D: New developments in soybeans.
Problem areas:
1. Reviewing and evaluating soybean production developments.
2. Reviewing and evaluating soybean marketing developments.

**Enterprise: Wheat**

Unit A: Growing wheat.
Problem areas:
1. Selecting and purchasing seed wheat.
2. Planting and fertilizing wheat.
3. Caring for the growing wheat crop.

Unit B: Harvesting, storing, and marketing wheat.
Problem areas:
1. Harvesting wheat.
2. Storing wheat.
3. Marketing wheat.

Unit C: New developments in wheat.
Problem areas:
1. Reviewing and evaluating wheat production developments.
2. Reviewing and evaluating wheat marketing developments.

**Enterprise: Forage crops**
Unit A: Establishing pastures.
Problem areas:
1. Selecting seed mixtures for pastures.
2. Establishing permanent pastures.
3. Establishing rotation pastures.

Unit B: Maintaining pastures.
Problem areas:
1. Maintaining permanent pastures.
2. Renovating permanent pastures.

Unit C: Growing hay.
Problem areas:
1. Selecting seed mixtures for hay crops.
2. Planting grasses and legumes for hay.
3. Controlling insects and diseases of hay crops.

Unit D: Harvesting, storing, and marketing hay.
Problem areas:
1. Harvesting and storing hay.
2. Marketing hay.
3. Harvesting and marketing seed for hay crops.

Unit E: Producing corn silage.
Problem areas:
1. Growing corn for silage.
2. Harvesting and storing corn for silage.

Unit F: Producing grass silage.
Problem areas:
1. Growing grasses and legumes for silage.
2. Harvesting and storing grasses and legumes for silage.

Unit G: New developments in forage crops.
Problem areas:
1. Reviewing and evaluating pasture production developments.
2. Reviewing and evaluating hay production and marketing developments.
3. Reviewing and evaluating corn silage production developments.

4. Reviewing and evaluating grass silage production developments.

**Enterprise: Soils and Fertilizers**

Problem areas:
1. Taking soil samples.
2. Making and interpreting soil tests.
3. Making and interpreting tissue tests and observing deficiency symptoms.
4. Selecting and buying fertilizers.
5. Preparing the soil for planting.
6. Maintaining the organic matter content of the soil.
7. Handling and storing manure.
8. Making a soil conservation plan.
9. Determining soil capacity.
10. Planning the long-time fertilizer program for the farm.
11. Reviewing and evaluating soil fertility and management developments.

**Enterprise: Agricultural Mechanics**

Unit A: Beginning agricultural mechanics.
Problem areas:
1. Getting acquainted with the agricultural lab.
2. Using safety practices in the agricultural lab.
3. Reconditioning and caring for hand tools.
4. Developing basic carpentry skills.
5. Developing the home shop.

Unit B: Advanced agricultural mechanics.
Problem areas:
1. Developing basic welding skills.
2. Developing advanced agricultural carpentry skills.
3. Using safety practices in the agricultural lab.
4. Improving the home shop.

Unit C: Agricultural machinery.
Problem areas:
1. Repairing and adjusting tillage equipment.
2. Repairing and adjusting planting equipment.
3. Repairing and adjusting spraying equipment and other specialized machines.
4. Repairing and adjusting harvesting and storing equipment.
5. Repairing and adjusting livestock equipment.
6. Repairing and adjusting feed handling equipment.

Unit D: Small agricultural engines.
  Problem areas:
  1. Operating small engines.
  2. Caring for and adjusting the fuel system components.
  3. Selecting fuels and lubricants.
  4. Caring for and adjusting the ignition system.
  5. Replacing valves, rings, and bearings.
  6. Trouble shooting.

Unit E: Agricultural equipment.
  Problem areas:
  1. Caring for and making adjustments in the electrical system.
  2. Caring for and making adjustments in the fuel system.
  3. Caring for and making adjustments to valves, rings, and bearings.
  4. Caring for the cooling system.
  5. Caring for tires.
  6. Caring for the suspension system and steering mechanism.
  7. Caring for the power train.
  8. Selecting fuels and lubricants.
  9. Operating large agricultural engines.
  10. Caring for and adjusting brakes.
  11. Caring for hydraulic systems.
  12. Trouble shooting.
  13. Operating equipment.

Unit F: Electricity.
  Problem areas:
  1. Wiring agricultural buildings.
  2. Caring for electric motors.
  3. Selecting electric motors.
  4. Using electrical controls.

Unit G: Metal work.
  Problem areas:
  1. Developing advanced welding skills.
  2. Developing cold metal work skills.
  3. Extending and maintaining agricultural plumbing.

Unit H: Agricultural buildings and construction.
  Problem areas:
  1. Using masonry in construction.
  2. Working with concrete.
  3. Repairing and improving agricultural buildings.
  4. Building fences.

Unit I: Building conservation structures.
Problem areas:
1. Constructing ponds.
2. Constructing grass waterways.
3. Constructing terraces.
4. Laying out contours.
5. Controlling gully erosion.
6. Surveying for conservation structures.

Unit J: Providing irrigation and drainage for the land.
Problem areas:
1. Constructing a land drainage system.
2. Installing an irrigation system.

Unit K: New developments in agricultural mechanics and engineering.
Problem areas:
1. Reviewing and evaluating soil conservation developments.
2. Reviewing and evaluating water management developments.

**Enterprise: Managing and Operating the Farm / Ranch / Agribusiness.**

Unit A: Keeping and using records.
Problem areas:
1. Keeping records.
2. Summarizing records.
3. Analyzing and using records.
4. Preparing income tax and social security reports.
5. Using microcomputers.

Unit B: Determining the production enterprises for the farm / ranch business.
Problem areas:
1. Taking an inventory of the farm / ranch business resources.
2. Planning the cropping system.
3. Planning the livestock system.
4. Considering recreation as a farm / ranch enterprise.

Unit C: Planning for sales and purchases.
Problem areas:
1. Planning the long-time marketing program.
2. Planning the long-time purchasing program.

Unit D: Planning for buildings and equipment for the business.
Problem areas:

1. Determining building needs.
2. Planning for the water supply.
3. Selecting machinery and power units.

Unit E: Financing the business.
Problem areas:
1. Identifying sources of credit for the business.
2. Determining the credit needs for the business.
3. Securing credit for the business.
4. Insuring the business.

Unit F: Understanding government programs and agricultural law.
Problem areas:
1. Developing leases and agreements.
2. Getting acquainted with agricultural laws.
3. Describing federal agricultural programs.

Unit G: Meeting personnel and personnel management needs.
Problem areas:
1. Determining management personnel needed.
2. Determining labor needed.
3. Securing and managing the labor force.

Unit H: Family finance.
Problem areas:
1. Providing family insurance.
2. Describing social security.
3. Planning a family investment program.
4. Planning the family budget.

**Enterprise: Rural Living**

Unit A: Understanding rural organizations and rural-urban relationships.
Problem areas:
1. Getting acquainted with agricultural and other rural organizations.
2. Improving rural-urban relationships.
3. Improving the rural community.

Unit B: Improving the farmstead.
Problem areas:
1. Planning for farm safety.
2. Controlling brush.
3. Controlling noxious and poisonous weeds.

Unit C: Improving the home.
Problem areas:

1. Planning for safety in the home.
2. Landscaping the home grounds.

Unit D: Growing food for home consumption.
Problem areas:
1. Planning the home garden.
2. Planting the home garden.
3. Caring for the home garden.
4. Harvesting garden produce.

**Enterprise: Natural Resources**

Unit A: Understanding the economic importance of renewable natural resources.
Problem Areas:
1. Identifying value of forests.
2. Identifying value of soil and water.
3. Identifying value of wildlife.
4. Identifying value of outdoor recreation.

Unit B: Exploring career opportunities.
Problem Areas:
1. Determining the career opportunities at various levels.
2. Determining the career opportunities by location.
3. Determining the educational opportunities beyond high school.

Unit C: Understanding tree growth.
Problem Areas:
1. Identifying the parts of a tree.
2. Determining the function of tree parts.
3. Determining how trees grow.

Unit D: Identifying tree species.
Problem Areas:
1. Determining the importance of tree identification.
2. Identifying trees by various features.

Unit E: Understanding forest ecology.
Problem areas:
1. Identifying environmental factors affecting the forest site.
2. Determining the site.
3. Describing the environmental factors in seed dissemination.
4. Determining the ecological factors in plant succession.
5. Determining stand competition factors and their effects.
6. Determining forest types.
7. Distinguishing between even-aged and uneven-aged stands.

Unit F: Managing the forest.
Problem areas:
1. Analyzing the stand.
2. Making intermediate cuttings.
3. Making harvest cuttings.

Unit G: Reproducing the forest.
Problem areas:
1. Preparing site.
2. Providing a source of seed.
3. Providing for natural reproduction.
4. Using artificial methods of reproduction.

Unit H: Measuring and marketing forest products.
Problem Areas:
1. Defining units of measurement.
2. Using log rules.
3. Measuring logs.
4. Estimating standing timber (board feet).
5. Estimating standing timber (cubic feet).
6. Determining products to be sold.
7. Selecting the markets.
8. Marketing by-products.

Unit I: Using maps and aerial photographs in environmental planning.
Problem areas:
1. Identifying maps and their use.
2. Using maps, scales, directions, and symbols.
3. Determining elevation and relief.
4. Using topographic maps.
5. Describing the fundamentals of aerial photographs.
6. Using aerial photographs.

Unit J: Identifying important wildlife species and their environmental preferences.
Problem Area:
1. Identifying fish species and their habitats.
2. Identifying bird species and their habitats.
3. Identifying animal species and their habitats.

Unit K: Surveying and leveling for conservation.
Problem Areas:
1. Determining the need for surveying and leveling.
2. Selecting surveying and leveling equipment.
3. Caring for and using equipment.
4. Surveying.
5. Making maps for conservation planning.

Unit L: Determining land capability and planning for its use.
   Problem areas:
   1. Determining the physical characteristics.
   2. Determining the best use.
   3. Using land capability maps and soil surveys.

Unit M: Protecting the forest.
   Problem areas:
   1. Protecting against forest fires.
   2. Protecting against insects and diseases.
   3. Controlling animals.
   4. Controlling vandalism.

Unit N: Harvesting forest products.
   Problem Areas:
   1. Selecting trees.
   2. Locating roads and operation sites.
   3. Felling.
   4. Limbing and bucking.
   5. Handling and skidding.
   6. Loading.
   7. Hauling.

Unit O: Processing forest products.
   Problem areas:
   1. Sawing lumber and other wood products.
   2. Processing pulp, particle, and fiber products.
   3. Manufacturing plywood and veneer.
   4. Processing post, poles, and pilings.
   5. Remanufacturing forest products.

Unit P: Grading lumber.
   Problem Areas:
   1. Determining reasons for grading.
   2. Describing the advantages to the industry.
   3. Exploring the opportunities available to lumber graders.
   4. Grading hardwood and softwood lumber.

Unit Q: Constructing ponds.
   Problem areas:
   1. Determining the value of ponds.
   2. Locating the ponds.
   3. Securing technical advice.
   4. Designing dam and spillway.
   5. Determining method of construction.
   6. Providing labor, equipment, and materials.
   7. Maintaining the pond.

### Enterprise: Growing and Marketing Christmas Trees

Problem areas:
1. Selecting site.
2. Selecting varieties.
3. Transplanting.
4. Planting.
5. Pruning and shaping.
6. Harvesting and grading.
7. Tying and marketing.

### Enterprise: Exploring Opportunities for Establishing a Recreation Business

Problem areas:
1. Using natural areas.
2. Using man-made areas.
3. Evaluating specific considerations.

### Enterprise: Fish Pond Ecology and Management

Problem areas:
1. Stocking.
2. Fertilizing.
3. Controlling weeds.
4. Controlling undesirable species.
5. Harvesting.
6. Maintaining population balance.

### Enterprise: Planning and Operating a Hunting and Fishing Preserve

Problem areas:
1. Describing game laws and regulations.
2. Planning general features.
3. Planning land use.
4. Providing guides.
5. Leasing.
6. Charging hunting, fishing, and trapping fees.
7. Providing boats.
8. Providing horses.
9. Releasing game birds and animals.
10. Retrapping.
11. Controlling diseases and predators.
12. Practicing safety.
13. Providing insurance.

**Enterprise: Wildlife Ecology and Management**

Problem areas:
1. Stimulating populations.
2. Using artificial propagation.
3. Providing effective regulations.
4. Using live trapping.
5. Tagging.
6. Controlling predators.
7. Controlling diseases.

**Enterprise: Determining the Feasibility of Conducting an Enterprise such as Forestry, Outdoor Recreation, Wildlife, and Urban Conservation**

Problem areas:
1. Determining the nature and scope of the enterprise.
2. Selecting the site.
3. Determining the resources available.
4. Estimating the available market.
5. Determining the interest, knowledge, and ability of the operator.
6. Appraising the needs of the community.
7. Determining the interest of the local community.
8. Evaluating the influence of other enterprises in the area.
9. Developing a preliminary and long-range plan.
10. Preparing a cost-return analysis.
11. Determining sources of information and assistance.

**Enterprise: Developing and Operating a Campground**

Problem areas:
1. Determining suitability of site.
2. Providing rest rooms.
3. Providing for refuse disposal.
4. Providing water supply.
5. Providing fuel and cooking facilities.
6. Maintaining the facilities.
7. Charging fees.
8. Providing insurance.
9. Providing for safety.

**Enterprise: Hiking and Riding Trails**

Unit A: Developing and operating hiking and riding trails.
Problem areas:
1. Locating.
2. Developing self-guiding trails.

## APPENDIX B

    3. Marking and clearing.
    4. Bridging.
    5. Developing rest areas.
    6. Providing for safety.
    7. Providing insurance.

Unit B: Planning and operating a riding stable.
    Problem areas:
    1. Planning the facility.
    2. Providing riding horses.
    3. Feeding and housing riding horses.
    4. Providing equipment.
    5. Providing food and drinks.
    6. Teaching horsemanship.
    7. Promoting horse shows.
    8. Charging fees.
    9. Providing for safety.
    10. Providing insurance.
    11. Maintaining the facility.

**Enterprise: Developing and Operating a Marina.**

    Problem areas:
    1. Planning the facility.
    2. Operating the marina.
    3. Providing for safety.
    4. Providing insurance.

**Enterprise: Developing and Operating a Vacation Farm**

    Problem areas:
    1. Determining the feasibility.
    2. Financing the vacation farm.
    3. Managing the business.
    4. Promoting the business.
    5. Providing for safety.
    6. Providing insurance.

**Enterprise: Apple Production**

Unit A: Developing the apple enterprise.
    Problem areas:
    1. Selecting and buying trees for planting.
    2. Planting the new orchard.
    3. Caring for the growing trees to bearing age.
    4. Grafting and budding.

Unit B: Controlling diseases and insects of apples.
    Problem areas:

1. Controlling insects of apples.
2. Preventing and controlling diseases of apples.

Unit C: Caring for the mature apple orchard.
Problem areas:
1. Pruning the mature apple tree.
2. Thinning fruit.
3. Cultivating and intercropping.
4. Fertilizing the orchard.
5. Renovating old orchards.

Unit D: Harvesting and marketing apples.
Problem areas:
1. Harvesting apples.
2. Storing apples.
3. Packing and marketing apples.

Unit E: New developments in apples.
Problem areas:
1. Reviewing and evaluating apple production developments.
2. Reviewing and evaluating apple marketing developments.

**Enterprise: Tomato Production**

Unit A: Planting tomatoes.
Problem areas:
1. Selecting and buying plants.
2. Growing tomato plants from seed.
3. Transplanting tomato plants.
4. Fertilizing tomatoes.
5. Contracting for tomatoes.

Unit B: Controlling diseases and pests of tomatoes.
Problem areas:
1. Controlling insects.
2. Preventing and controlling diseases.
3. Controlling weeds in tomatoes.

Unit C: Harvesting and marketing tomatoes.
Problem areas:
1. Picking tomatoes.
2. Hauling tomatoes to market.
3. Marketing tomatoes.

Unit D: New developments in tomatoes.
Problem areas:
1. Reviewing and evaluating tomato production developments.

## APPENDIX B

2. Reviewing and evaluating tomato marketing developments.

**Enterprise: Horses**

Unit A: Light horses.
Problem areas:
1. Feeding and caring for light horses.
2. Training and handling light horses.
3. Selecting and buying light horses.

**Enterprise: Rice[1]**

Unit A: Selecting and buying rice seeds.
Problem areas:
1. Selecting the rice seeds.
2. Chemically treating and storing of selected seeds.
3. Buying good rice seeds.

Unit B: Preparing the seedbed, sowing, and caring for rice seedlings.
Problem areas:
1. Preparing the seedbeds.
2. Sowing the rice seeds.
3. Caring for the growing seedlings until transplanting.

Unit C: Preparing the rice field and transplanting seedlings.
Problem areas:
1. Plowing the land and preparing the dikes.
2. Harrowing and leveling the rice field.
3. Irrigating the rice field.
4. Applying fertilizer at planting time.
5. Transplanting the rice seedlings.

Unit D: Caring for the growing crop.
Problem areas:
1. Irrigating the growing crop.
2. Applying fertilizer to the growing crop.
3. Controlling weeds in the growing crop.
4. Preventing and controlling pests in the growing crop.
5. Preventing and controlling disease in rice.

Unit E: Harvesting, threshing, and milling rice.
Problem areas:
1. Harvesting the rice crop.
2. Threshing the rice crop.
3. Milling rice.

---

[1] Developed with Vincente A. Quiton for the Philippines.

Unit F: Storing and marketing the rice crop.
  Problem areas:
  1. Storing the rice crop for market.
  2. Marketing the rice crop.

Unit G: New developments in rice.
  Problem areas:
  1. Reviewing and evaluating research findings in rice production.
  2. Reviewing and evaluating harvesting, storing, and milling developments in rice.
  3. Reviewing and evaluating marketing developments in rice.

**Enterprise: Ornamental Horticulture (For instructional purposes, select plants appropriate for class and community.)**

Unit A: Propagating horticulture plants.
  Problem areas:
  1. Selecting tools and equipment.
  2. Producing plants from seed.
  3. Reading labels.
  4. Germinating seed.
  5. Planting the seed.
  6. Caring for the seedlings.

Unit B: Propagating plants from cuttings.
  Problem areas:
  1. Identifying types of cuttings.
  2. Describing factors which affect success of cuttings.
  3. Making various cuttings.
  4. Preparing cuttings for rootings.
  5. Placing cuttings in rooting media.
  6. Caring for the cuttings.
  7. Selecting tools and equipment.

Unit C: Propagating plants by layering.
  Problem areas:
  1. Describing differences in layering and cuttings.
  2. Identifying kinds of layering.
  3. Caring for new plants before planting.

Unit D: Propagating plants by budding and grafting.
  Problem areas:
  1. Describing methods of budding.
  2. Grafting of plants.
  3. Caring for grafted plants.
  4. Selecting tools and equipment.

Unit E: Constructing plant propagation containers and equipment.
Problem areas:
1. Constructing flats and planting boards.
2. Employing safety precautions.

Unit F: Growing horticulture plants.
Problem areas:
1. Caring for plants.
2. Transplanting plants.

Unit G: Using soil and other plant-growing media.
Problem areas:
1. Using soil media in growing plants.
2. Identifying soil conditioners.
3. Preparing soil mixtures.
4. Using soil mulches in growing plants.
5. Controlling erosion by using plant-growing media.

Unit H: Constructing, maintaining, and using plant-growing structures.
Problem areas:
1. Identifying and using plant-growing structures.
2. Identifying and using types of heat.
3. Identifying and using air-cooling systems.
4. Operating plant-growing equipment.

**Enterprise: Landscaping**

Unit A: Establishing lawns and turf.
Problem areas:
1. Selecting tools and equipment.
2. Preparing the seed bed.
3. Describing the differences in lawn grasses.
4. Seeding lawns.
5. Transplanting turf.
6. Transplanting shrubs.
7. Caring for and maintaining tools and equipment.

Unit B: Maintaining lawns and turfs.
Problem areas:
1. Selecting tools and equipment.
2. Mowing the lawn.
3. Fertilizing the lawn.
4. Aerating the lawn.
5. Watering the lawn.
6. Controlling pests of lawn.
7. Caring for and maintaining tools and equipment.

Unit C: Repairing or renovating lawns.
Problem areas:
1. Identifying causes of deterioration.
2. Selecting tools and equipment.
3. Repairing or renovating lawns.
4. Caring for and maintaining tools and equipment.

**Enterprise: Gardening**

Problem areas:
1. Describing needs and values of a home garden.
2. Selecting the location of a garden.
3. Selecting crops, varieties, seeds, and plants.
4. Estimating seed and fertilizer.
5. Preparing the soil.
6. Fertilizing the soil.
7. Planting the seeds.
8. Controlling the pests.
9. Watering the plants.
10. Harvesting the vegetables.
11. Constructing and using hotbeds and cold frames.

**Enterprise: Home Beautification**

Problem areas:
1. Maintaining driveways.
2. Maintaining walks.
3. Maintaining the house and general repair.

**Enterprise: Small Animals**

Unit A: Feeding animals.
Problem areas:
1. Weighing and measuring feed.
2. Reading labels and directions.
3. Feeding the animals.
4. Watering the animals.
5. Keeping records.

Unit B: Caring for animals.
Problem areas:
1. Cleaning pens.
2. Disinfecting pens.
3. Bedding animals.
4. Recognizing animal pests and diseases.
5. Exercising animals.
6. Grooming animals.

# AGRICULTURAL BUSINESS COURSE

## Part I: General Agricultural Business and Orientation

Note: It is assumed that all students enrolled in the agricultural business course will have taken some other agricultural courses as prerequisites. Thus the students will have had instruction on agricultural occupations.

Unit A: Developing a placement-employment occupational experience program in agriculture.
Problem areas:
1. Getting acquainted with placement-employment programs.
2. Arranging for the placement-employment program in an agricultural business.
3. Keeping placement-employment program records.
4. Managing personal income from the placement-employment program.

Unit B: Getting a job.
Problem areas:
1. Knowing yourself.
2. Describing work laws.
3. Locating agricultural job opportunities.
4. Applying for a job and being interviewed.

Unit C: Studying agricultural businesses.
Problem areas:
1. Determining how agricultural firms are organized to do business.
2. Studying the agricultural training station business and the occupations in it.
3. Getting acquainted with the business responsibilities of the employer.
4. Getting acquainted with the business responsibilities of the employee.
5. Determining credit risks.

Unit D: Buying and handling agricultural products.
Problem areas:
1. Buying agricultural products for resale.
2. Transporting and handling agricultural products.

Unit E: Advertising and promoting the business.
Problem areas:
1. Pricing agricultural products and services.
2. Promoting agricultural products.
3. Developing good customer relationships.

Unit F: Selling agricultural products and services.
Problem areas:
1. Developing selling techniques for agricultural products.
2. Developing selling techniques for agricultural services.
3. Operating sales business machines and equipment.
4. Writing up the sales slip.

Unit G: Financing the agricultural business.
Problem areas:
1. Identifying sources of credit.
2. Determining credit needs.
3. Securing credit for the business.

Unit H: Keeping and using agricultural business records.
Problem areas:
1. Keeping sales and purchasing records.
2. Keeping personnel records.
3. Taking inventories.
4. Keeping records of business costs.
5. Keeping records of services rendered.
6. Using records to improve the business.

Unit I: Providing for safety in the business.
Problem areas:
1. Handling problems of customers.
2. Handling accidents to personnel or customers.
3. Managing customers in case of fire or other hazards.
4. Planning for safety in the business.

## Part II. Developing Knowledge and Skills Needed in Agricultural Placement

The particular areas involved here will depend on the kinds of agricultural businesses in which the students are obtaining placement-employment experiences. The students will need to study intensively about all of the products and services available in their training station business.

Much of the study of agriculture will be, of necessity, individual in nature. However, as much as possible of the agricultural knowledge and skill needed should be provided through group instruction.

As a starting place for the study of agriculture, the teacher can select any of the "new developments" units taught in the other agriculture courses. All of these new developments units should be taught to the entire agricultural business class sometime during the year.

## APPENDIX B

If other areas in agriculture common to the entire class, or to several of the class, can be identified, these areas should be taught on a group instruction basis.

A list of some of the agriculture units which may be taught, in addition to those listed relating to the agricultural, is provided here for discussion purposes only.

Unit A: Knowing your agricultural products and services.
    Problem areas:
    1. Getting acquainted with your agricultural products.
    2. Identifying your agricultural services.

Unit B: Continuing participation in the FFA.
    Problem areas:
    1. Evaluating and replanning programs of work.
    2. Improving leadership ability.

Unit C: New developments in swine.
    Problem areas:
    1. Reviewing swine production developments.
    2. Reviewing swine marketing developments.

Unit D: New developments in beef.
    Problem areas:
    1. Reviewing beef production developments.
    2. Reviewing beef marketing developments.

Unit E: New developments in dairying.
    Problem areas:
    1. Reviewing dairy production developments.
    2. Reviewing dairy marketing developments.

Unit F: New developments in sheep.
    Problem areas:
    1. Reviewing sheep production developments.
    2. Reviewing sheep marketing developments.

Unit G: New developments in poultry.
    Problem areas:
    1. Reviewing poultry production developments.
    2. Reviewing poultry marketing developments.

Unit H: New developments in corn.
    Problem areas:
    1. Reviewing corn production developments.
    2. Reviewing corn marketing developments.

Unit I: New development in oats.
    Problem areas:

1. Reviewing oats production developments.
2. Reviewing oats marketing developments.

Unit J: New developments in soybeans.
Problem areas:
1. Reviewing soybean production developments.
2. Reviewing soybean marketing developments.

Unit K: New developments in wheat.
Problem areas:
1. Reviewing wheat production developments.
2. Reviewing wheat marketing developments.

Unit L: New developments in forage crops.
Problem areas:
1. Reviewing pasture production developments.
2. Reviewing hay production and marketing developments.
3. Reviewing corn silage production developments.
4. Reviewing grass silage production developments.

Unit M: New developments in soils and agricultural mechanization.
Problem areas:
1. Reviewing soil fertility developments.
2. Reviewing soil conservation developments.
3. Reviewing water management developments.

Unit N: The rural community.
Problem areas:
1. Getting acquainted with agricultural and rural organizations.
2. Improving rural-urban relationships.

Unit O: Conserving wildlife.
Problem areas:
1. Conserving game animals.
2. Conserving game birds.
3. Conserving fish.
4. Conserving wild plants.

Unit P: New developments in apples.
Problem areas:
1. Reviewing apple production developments.
2. Reviewing apple marketing developments.

Unit Q: New developments in tomatoes.
Problem areas:
1. Reviewing tomato production developments.
2. Reviewing tomato marketing developments.

## ADJUSTMENTS TO LOCAL SITUATIONS

It should always be remembered that adjustments in the analysis of enterprises into units and problem areas are made according to local conditions. Some teachers may wish to reduce an enterprise to one or two problem areas, or even to combine several enterprises, because of the limited importance of those enterprises locally and in the state. The following are examples of the kinds of adjustments teachers may wish to consider:

**Enterprise: Small Grains**

Unit A: Oats.
Problem areas:
1. Growing oats.
2. Harvesting, storing, and marketing oats.

Unit B: Wheat.
Problem areas:
1. Growing wheat.
2. Harvesting, storing, and marketing wheat.

**Enterprise: Personal Development**

Problem area: Managing money.

If the general principles of course building are applied, the resulting analysis of content will be what is needed for the program.

# APPENDIX

## C

# Using a Problem-Solving Approach in Other Teaching Situations

THE PROBLEM-SOLVING APPROACH PRESENTED in this book can be applied in any teaching situation or for any topic. Many times when problem solving is mentioned as a way to teach topics other than vocational subjects, the immediate response made is that it won't work. The plans included in this section illustrate that problem solving can be used in any teaching situation.

The first two plans are examples of a problem-solving approach that has been used in an undergraduate methods course. The remaining plans focus upon instructional topics in an academic area.

### PLAN NUMBER ONE

Part I. Pre-planning Analysis
    A. Enterprise: Teaching techniques (in a methods course)
    B. Unit: Using different teaching techniques
    C. Problem area: Developing effective questioning
    D. Analysis of the teaching situation:
        1. Students will be student teaching the following quarter.
        2. Students have had educational psychology courses.
        3. Students will be expected to lead many discussions while teaching.
    E. Teaching / learning objectives: All class members should develop the ability to:
        1. Stimulate and motivate students by asking questions.
        2. Ask questions in a correct manner.
        3. Use questioning to determine knowledge possessed by students.
        4. Use questioning to increase students' understanding of technical information.

**Part II. The Teaching Plan**
  A. Unit: Using teaching techniques
  B. Problem area: Developing effective questioning
  C. Interest approach (leading discussion questions)
   1. How do you communicate with others (speech, sight, symbols, etc.)?
   2. How might you proceed to stimulate students?
   3. Why do you ask questions?
   4. How do you ask questions?
  D. Anticipated group objectives: (Why must you be knowledgeable in the area of questioning?)
   1. To increase teaching effectiveness.
   2. To become a master teacher.
  E. Anticipated problems and concerns:
   (What questions must you be able to answer in order to use questioning effectively?)
   1. What are the students' names?
   2. What are the different types of questions?
   3. What do you want the students to learn?
   4. How do you ask questions to promote thinking in the students?
   5. How do you state questions clearly?
   6. What procedure should be followed in asking questions?
   7. What is questioning?
  F. Steps in solving problems / concerns: (Decide which one to cover first.)
  G. Evaluation and application:
   1. Evaluation
     a. Listen to class members as they use questioning.
     b. Check lesson plans for questions.
     c. Observe the use of questioning during student teaching.
   2. Application
     a. Evaluate one another on the proper use of questioning.
     b. Develop and use good questions in lesson plans.
     c. Use questioning while teaching a mini lesson in a methods course and in student teaching.

**Part III. Teaching Resources**
  1. *Teaching Agriculture Through Problem Solving.*
  2. Mimeo, "Elements of Good Questioning."
  3. File folder on questioning.
  4. Chalkboard.
  5. Mimeo, "The Teacher as a Professional Question Maker."
  6. Mimeo, "Study Questions."

## PLAN NUMBER TWO

**Part I. Pre-planning Analysis**

  A. Enterprise: Preparing for teaching
  B. Problem area: Developing lesson plans
  C. Analysis of the teaching situation:
   1. Students will be student teaching next quarter.
   2. Students have had educational psychology courses.
   3. Students will be expected to prepare lesson plans.
  D. Teaching / learning objectives: The students will be able to:
   1. Recognize the value of sound lesson planning.
   2. List the major components of a lesson plan.
   3. Describe the purpose of each of the major components.
   4. Use the lesson plan in the improvement of instruction in their local schools.

**Part II. The Teaching Plan**

  A. Enterprise: Preparing for teaching
  B. Problem area: Developing lesson plans
  C. Interest approach: Using samples of water pollution, lead a discussion on what would be the bare essentials of a lesson plan if you were developing a lesson plan on controlling water pollution. (List the input from the class on the chalkboard.)
  D. Anticipated group objectives: (Why must you be knowledgeable in the procedure to use in developing lesson plans?)
   1. To be prepared for teaching a specific unit or problem area.
   2. To increase teaching effectiveness.
   3. To become a master teacher.
  E. Anticipated problems and concerns: (What questions must you be able to answer about lesson planning in order to develop an effective lesson plan?)
   1. What are the components of a lesson plan?
   2. What are the purposes of each of the lesson plan components?
   3. How can the lesson plan format be used in a local teaching situation?
   4. Why develop lesson plans?
  F. Steps in solving problems / concerns. (Lead the students in a logical order to follow in developing lesson plans.)
  G. Evaluation and application:
   1. Evaluation
     a. Paper / pencil test.
     b. Lesson plans developed follow recommended format.

2. Plans for application
   a. Make lesson plans for mini teaching in a methods course.
   b. Make lesson plans for student teaching.

**Part III. Teaching Resources**
1. Transparency on lesson plan format.
2. Lesson plan format.
3. Transparency of teaching calendar format.
4. Curriculum guide for Ag. Science and Mechanics I & II.
5. Samples of water depicting pollution.
6. Sample lesson plans on selecting lubricants and using the futures market.
7. Overhead projector, chalkboard, colored chalk, felt tip pens, and 3 x 5 cards.

## PLANS FOR OTHER AREAS

In a small high school in Homer, Illinois, vocational agriculture instructor Floyd Fuller[1] was also required to teach a chemistry course. As a result of his experiences in using the problem-solving approach presented in this book for his vocational agriculture classes, he decided to try the same procedures in his chemistry class. His first plans are included here as a matter of interest, together with the comments made by his pupils when he evaluated his experience.

### Plan Number One (three days)

**Part I. Pre-planning Analysis**
   A. Enterprise: Photography
   B. Problem area: The camera
   C. Analysis of the situation:
   For several years photography has been gaining in popularity as a hobby. This is probably due to more people having more leisure time to spend and to the fact that it is interesting. Almost everyone in our locality takes some pictures each year. The students in the chemistry class are at the age when this is particularly interesting to them. In order for them to obtain the most knowledge from this subject, they should first learn a few things about the camera.

---
[1] Floyd Fuller, former teacher of vocational agriculture, Homer, Illinois.

## APPENDIX C

**Part II. The Teaching Plan**

A. Teacher objectives: To develop:
   1. A knowledge of the single convex lens and the anastigmatic lens.
   2. An ability to distinguish between the above lenses.
   3. An understanding of the focal length of the lens.
   4. A knowledge of the diaphragm and its use.
   5. An understanding of the relative aperture or "F" value.
   6. A knowledge of shutters.
   7. An understanding of the shutter speed settings.
   8. An understanding of the differences between box cameras, candid cameras, and minature cameras.
B. Teaching procedures
   1. Interest approach (leading questions)
      a. How many of you have a camera at home?
      b. How many of you have used a camera?
      c. How many parts of a camera can you name?
      d. What do these parts do?
   2. Study a camera in class.
   3. Take a field trip to learn how to operate a camera and to take pictures for further study.
C. Pupil objectives:
   1. To learn how to operate a camera.
   2. To learn the different parts of a camera.
D. Anticipated problems of the pupils
   1. What is an anastigmatic lens?
   2. What is the difference in lenses?
   3. What is the focal length of a lens?
   4. What is the diaphragm and how does it work?
   5. What does "F" mean?
   6. How does a shutter work?
   7. What is the difference between box cameras, candid cameras, and miniature cameras?
E. Evaluation and application
   1. Approved practices
      a. Use the best lens for the picture to be taken.
      b. Adjust the "F" setting to the light and type of subject.
      c. Adjust the shutter speed according to the rate of motion in the picture.
      d. Purchase a camera to serve your specific needs.
   2. Test at end of unit.
   3. Actual practice—at least some of the students should become interested enough in photography to buy their own cameras and use them regularly.

Part III. Teaching Resources
1. Camera.
2. Demonstration with FFA camera.
3. Field trip with instructions on the use of a camera.

## Plan Number Two (three days)

Part I. Pre-planning Analysis
A. Enterprise: Photography
B. Problem area: Developing the negative
C. Analysis of the teaching situation:
The chemistry class is interested in learning to develop a negative. They are mostly interested in the subject as a hobby, but they should learn the process from the chemical standpoint as well since this is a chemistry class.

Part II. The Teaching Plan
A. Teacher objectives: To develop:
1. A knowledge of the chemical reaction which takes place when a film is exposed.
$$2\,AgBr + light \quad Ag_2Br + Br \bullet gelatin$$
2. A knowledge of the need for using darkroom bulbs when working with exposed film.
3. A knowledge of the four steps in the developing process.
4. An ability to arrange the developing pans in the correct order.
5. An ability to mix the solutions needed.
6. An ability to develop a negative.
7. The knowledge that the developer further oxidizes the $Ag_2Br$ to free silver.
8. The knowledge that the hypo (sodium thiosulfate) removes the unaffected $AgBr - Ag_2Br$ to keep the film from turning black upon exposure to light.
B. Teaching procedures
1. Interest approach (leading questions)
a. Have any of you ever developed a negative?
b. Have any of you seen a negative developed?
c. How do you suppose a negative is developed?
d. Would you like to develop a negative?
e. What do you suppose happens in developing a negative?
f. What do we need to know, or need to know more about, to develop a negative?

APPENDIX C 303

    C. Pupil objectives
        1. To learn how to develop a negative.
        2. To learn what is needed to develop a negative.
        3. To learn what happens when a negative is developed.
    D. Anticipated problems of the pupils
        1. How do you develop a negative?
        2. What does a darkroom bulb look like?
        3. Why do we need to develop a negative?
        4. What are the steps in developing a negative?
        5. What are the solutions we need?
        6. What do these solutions do?
    E. Evaluation and application
        1. Approved practices
            a. Mix solutions according to directions.
            b. Arrange pans of solutions in handy order.
            c. Use a darkroom lighted with a darkroom bulb.
            d. Develop one negative.
        2. Test on unit.
        3. Actual practice—develop one negative apiece in the laboratory.

**Part III. Teaching Resources**

    1. Film.
    2. Developing Supplies.

## Plan Number Three (three days)

### Part II. The Teaching Plan

    A. Enterprise: Photography
    B. Printing the positive
    C. Analysis of the teaching situation:
    The chemistry class is studying the developing process. The students have taken pictures and developed the negatives. Now they are ready to print the positives and learn the process behind it.

### Part II. The Teaching Plan

    A. Teacher objectives: To develop:
        1. An understanding of the printing process.
        2. A knowledge of the chemical solutions used.
        3. A knowledge of the tray arrangement and the solutions in each.
        4. An understanding of the print paper.
        6. An understanding of the chemistry of printing.

B. Teaching procedure:
   1. Interest approach (Leading questions)
      a. How many of you have printed a negative?
      b. How many of you have seen a positive printed?
      c. How do you suppose a positive is printed?
      d. What do you suppose happens when you print a positive?
      e. Would you like to print a positive?
      f. What do we need to know or need to know more about to print a positive?
C. Pupil objectives
   1. To learn to print a positive.
   2. To learn what is needed to print a positive.
   3. To learn what happens chemically when a positive is printed.
D. Anticipated problems
   1. How do you print a positive?
   2. What solutions are necessary in printing a positive?
   3. How are the trays arranged?
   4. What does print paper look like?
   5. What is a print box and how is it used?
   6. What happens chemically when a positive is printed?
E. Evaluation and application
   1. Approved practices
      a. Prepare a darkroom for printing.
      b. Mix solutions according to directions.
      c. Arrange solutions in a handy order.
      d. Print a positive.
   2. Test on unit.
   3. Actual practice—print the positives of the negative developed in class.

**Part III. Teaching Resources**
   1. Appropriate solutions.
   2. Positive.
   3. Needed supplies.

After using the problem-solving method of teaching the chemistry of photography in a chemistry class of 13 students, Mr. Fuller asked for a written evaluation of the method from the students. After a slight review of the method the following questions were asked: "Do you like this method of studying chemistry? Why or why not?" Eleven students were present and these are their answers.

## APPENDIX C

Yes, I like it better in the last week. The reason is that we haven't been rushed as much and have had more time to discuss things. It seems more interesting the way we now study by writing the questions on the board and then discussing them. It is easier to answer the questions than like we used to do.

Yes, because I think I am getting more out of it and it is also an interesting subject.

Yes, I believe that we are getting more out of it by looking up the questions for ourselves. Another thing that will help if you write out the questions at the end of each one.

Yes, because I just like what we've been studying and enjoy it more.

Yes, because you have the question in writing and we answer them in class. I think it will help us in our semester test and also 6 weeks test. I think we should study more ways of photography and study or experiment more.

Yes, because I can understand it better. It makes it more interesting, and more direct way of studying.

Yes, because I think I am getting more out of it and I know what we are talking about when we discuss it in class, I feel free to talk in class without making a mistake and giving the wrong answer.

Yes, because by having the questions to look up it helps you more in knowing what you do need to know to develop film. By having the labs even though they weren't all in the lab book, I think we have gotten more out of it.

I like the way we are doing it a lot better than what we did before. Now we know just what we should know. When we do the questions before, you know what to look for in reading your book.

Yes, because I have learned more and I can understand it better and it is much more interesting.

Yes. It would help if I would take my book home, but I don't have time to study at home. I could study at school if I could concentrate in my study periods. I still would like to have chemistry the second period.

These answers were taken directly from the students' papers. Although the subject studied was one which gets much natural interest, it is interesting to note how many of the students found that it was easier to study the subject when the problem-solving method was used.

# APPENDIX D

## APPENDIX D

# Using and Developing "Principles" in Teaching Agriculture

AN AGRICULTURAL PRINCIPLE IS A GENERALIZATION which provides a basis for reasoning, a basis for making decisions, or a guide for conduct or procedure within an identified area of agricultural content or knowledge. Thus, there are "principles of corn production, some of which have application to corn only and others which have application to other crops. In other words, principles must be described as being *principles of something* within limits which are either stated or understood.

An example of an agricultural principle might be a "principle of weed control in corn" stated as follows:

> The best control of weeds in corn by the use of herbicides is achieved by application of the herbicide when the weeds are still very small.

This statement is a "generalization" in terms of its application to corn production and is, therefore, a "principle" of weed control in corn.

The principle, with the "in corn" words deleted, could also be a principle of weed control with application to weed control in most crops. Thus, the principle is a principle within a defined area of knowledge, the limits being either stated (as for corn) or understood (as relating to crops when the words "in corn" are deleted). Some agricultural principles may be limited to one kind of crop or livestock; other agricultural principles may be as broad as a biological or a physics principle. Each area of knowledge has its own set of principles, some of which may duplicate or overlap the principles of another area of knowledge. The overlapping and duplication are unifying forces among areas of knowledge.

Some efforts have been made to use the principles of an area of knowledge as the basis for curriculum construction. Principles do not serve well as the basis for organization of content. Principles have, by definition, "general" application. The illustrations of the general application of a principle could involve small parts of many different agricultural enterprises, units, and problem areas. These small parts of the many enterprises would have meaning in relation to the principle but not in relation to each other. In effect, small parts of the "nongeneralizable" content in agriculture would be extracted from their present locations as items closely tied to the other items around them and would be placed in a new location in which they would have a close tie only with the principle involved. This would make it very difficult and confusing, if not impossible, to teach students in an organized manner the specific content details needed for the application of principles to broad problem situations as found in the occupational field of agriculture. Because principles are "general" in application, it is the principle which should be used as the "peripetetic" element in the organization of content for instruction. The principle will be strengthened, rather than destroyed, by using or identifying it in many parts of the course of study. The structure of a particular area of knowledge would, on the other hand, be destroyed by an attempt to organize it on the basis of principles.

To use "principles" effectively in teaching agriculture, the principles need to be identified in relation to the presently recognized units of content as described in this book. Whenever an identified principle can be exposed to the student as a principle, this should be done. Eventually, students will have been exposed to the same principle several times, and they will begin to understand it as a principle having general application.

To make use of "principles" in teaching, the teacher needs to identify as many principles as possible and then relate them (by recording them on teaching plans, in source units, or on the course of study) to each unit of a course or courses where the principle has application. As additional principles are identified, they should be added to the proper lists. As the teacher identifies and uses principles in teaching in relation to the various units, a new richness of meaning and unity may be added to the total agricultural instructional program by the repeated use of the same principle in several contexts.

The following are some illustrations of the kind of statement of principles that will be developed as the teacher begins to make use of the "principles concept" in teaching. Not all of the statements may be true for all areas, since they were developed on the basis of limited knowledge of a small area. They should be used, therefore, only to provide a basis for understanding the "principles concept" and *not* as statements of principles to be used unless and until checked out by expert teachers for the area in which their use is desired.

**Enterprise: Corn**

Unit: Caring for the growing corn crop.

Problem areas:
1. Controlling weeds in the growing corn.
2. Controlling insects and pests in the growing crop.
3. Preventing and controlling diseases in corn.
4. Applying supplemental fertilizers.

*Principles of problem area #1:*

1. Proper timing of the use of a weed control method is important for effective weed control.
2. The degree of maturity of the weeds affects the selection of the weed control method used.
3. Plants may absorb herbicides through their leaves or through their roots.
4. To destroy weeds, the growth potential of the weeds must be destroyed.
5. Weeds compete with crops for nutrients and sunlight.
6. The rate of application of a herbicide for weed control varies with the kind of weed, the kind of crop plant, and the growing environment.
7. A rough surface on the seedbed helps control weeds.
8. Cultivation beyond that necessary to achieve good weed control may result in reduced yields.
9. Rotary hoeing should be done when weeds and crop plants are small.
10. The efficiency and effectiveness of herbicides are influenced by soil type, soil organic matter content, and soil micro-organisms.
11. Herbicides are decomposed most rapidly in soils with high organic matter content together with favorable moisture and heat conditions.
12. Herbicides are most effective in sandy soils, in sandy loams, and in loams with considerable sand.
13. Sunlight rapidly decomposes exposed herbicides.

14. The more permeable soils require larger applications of herbicides per acre than do the less permeable soils.
15. Some plants are able to dissolve certain herbicides and so remain unaffected by them.
16. Some herbicides may be applied in combination with fertilizer solutions.
17. The best control of weeds in corn from the use of herbicides is achieved by application of the herbicides before the weeds become large.
18. Flame weeding may be used only with crops more resistant to the heat of the flames than are the weeds.
19. There are several methods by which weeds can be controlled effectively.

*Principles of problem area #2:*

1. Plants differ in susceptibility and response to insect infestation.
2. Insecticides vary regarding selectivity in their effect on insects.
3. An insect may be most easily controlled at a particular time or point in its life cycle.
4. The cost of insect control beyond a certain level may exceed the economic gain from increased yield or improved quality.
5. The feeding method of an insect helps determine the best control measure to use.
6. Insects have natural enemies which may be used in controlling them.
7. Insects may be controlled by eliminating or destroying a part or all of their habitat.
8. Several insect pests may be controlled by the use of one kind of treatment or control measure.
9. A variety of control measures may be needed to control the insects which attack a single crop.
10. Using control measures at planting time is an effective means for reducing soil insect damage.
11. Insects may attack any part of a plant.
12. There is a variety of methods of value in controlling insect pests.
13. Insects may develop resistance to various kinds of controls.
14. Insects adapt to a changing environment.

*Principles of problem area #3:*

1. Disease prevention is less costly than treatment after the disease has made a successful attack.

## APPENDIX D

2. Every crop is subject to some diseases.
3. Plant diseases may be prevented or controlled by controlling intermediate hosts or the spreading agent.
4. Plant diseases may have any one of several causes—fungi, bacteria, viruses, nematodes, shortages of nutrients, weather, etc.
5. Transferring plants to new areas may expose them to new diseases to which the plant is not immune.
6. The development of new varieties of a plant may result in the new variety being susceptible to diseases to which the older varieties were resistant.
7. Some diseases may be controlled by cultural practices.
8. Insects or other causes may weaken plants and make the plants more susceptible to diseases.
9. The early recognition of the symptoms of disease is important to successful treatment.
10. The cost of disease prevention and control beyond a certain level may exceed the economic gain from increased yield or improved quality.
11. The use of resistant varieties is the most satisfactory method of controlling crop diseases.
12. Diseases are most easily controlled at particular stages in the development of the disease.

*Principles of problem area #4:*

1. As the maximum yield under optimum conditions is approached, increases in the level of nutrients may bring about such small increases in yields as to be uneconomical.
2. Plants use varying amounts of many nutrients from the soil.
3. Fertilizer should be applied only if other factors appear favorable for crop use of the fertilizer.
4. A deficiency of one nutrient needed for plant growth may limit the utilization of the other nutrients.
5. The returns from the use of fertilizer are greatest if all other factors affecting yield are at the most favorable level.
6. The soil nutrient supply is depleted by crop removal.
7. Many crops have a limited ability to absorb certain soil nutrients when the plant is young.
8. Improper placement of fertilizer may result in a high salt placement in the root zone.
9. Some forms of fertilizer are more easily leached from the soil than are other forms.

10. Phosphorus becomes less available with increasing soil acidity.
11. Most plant nutrient elements are more readily soluble at pH 6.5 than at a lower pH.
12. Most favorable soil microbes are more active at pH 6.5 than at a lower pH.
13. Some fertilizer elements may be applied as dry fertilizer, as sprays on foliage or the soil, or as gases.
14. Yields of crops vary with the amount of plant nutrients available in the soil.
15. The amount of fertilizer needed varies with the type of soil, the cropping history, and the yield desired.
16. Fertilizers vary in the kind and amount of available nutrients they contain.

# APPENDIX

# E

# Process Outline for Giving a Demonstration

THE TEACHER SHOULD TRY A "DRY RUN" on any demonstration prior to actually giving it. The "dry run" should follow the steps to be used in the actual demonstration.

1. *Orient the students to the demonstration.*—Explain what is to be demonstrated and how it relates to the instructional program. The purposes of the demonstration should be discussed.

2. *Show the students, if possible, what the demonstration is to produce or achieve.*—Having the finished product available for inspection will make it easier for the students to understand the demonstration.

3. *Show and describe the equipment and materials to be used.*—The group can be asked to name and describe equipment and materials needed with the teacher producing the items as they are named. The teacher can finish by showing items not named by the group.

4. *Emphasize safety.*—If goggles are required, students and teacher should be wearing them. The teacher should point out steps where accidents may occur and emphasize safe work habits at all times.

5. *Give the demonstration.*—Each step and important point should be identified and listed. Care must be taken to show and explain each step in a way students can see and understand. To the extent possible, the students can be asked to discuss the demonstration as it is being given. If additional time is available, related information may be injected into the procedures by the teacher. The amount of time to

be used in this way should be estimated during the "dry run" so that appropriate preparation can be made.

6. *Summarize as needed.*—Depending on the situation and teacher objectives, the teacher may summarize, a student may be called on to perform the demonstration, or the entire group may be directed to perform the activity demonstrated.

# Bibliography

# ▶ Bibliography ◀

1. Aiken, Wilford M. "Some Implications of the Eight-Year Study for All High Schools and Colleges." *North Central Association Quarterly*, Vol. 17, pp. 274-280, 1943.
2. Aiken, Wilford M. *The Story of the Eight-Year Study*. New York: Harper and Brothers, 1952.
3. *An Analysis of Standards for Teacher Education Programs in Agriculture / Agribusiness*. AATEA Ad Hoc Committee Report on the Legitimization of Program Standards and Guidelines, 1977.
4. *An Assessment of the Pedagogical Skills Taught to Agricultural Education Undergraduates*. AATEA Ad Hoc Committee Report on Teaching Techniques. Atlantic City, New Jersey: American Vocational Association Convention, 1977.
5. Bagford, Lawrence, Anthony S. Jones, and Edward Wallen. *Strategies in Education Explained*. Urbana, Ohio: Karlyn Publishing & Consulting, 1977.
6. Barlow, Melvin L., ed. *The Philosophy for Quality Vocational Education Programs*. Washington, D.C.: The Fourth Yearbook of the American Vocational Association, 1974.
7. Binkley, Harold R., and Rodney W. Tullock. *Teaching Vocational Agriculture / Agribusiness*. Danville, Illinois: The Interstate Printers & Publishers, Inc., 1981.
8. Bloom, Benjamin S. *Taxonomy of Educational Objectives, Handbook I: Cognitive Domain*. New York: David McKay, Inc., 1956.
9. Brownell, William A. "Problem Solving." *The Psychology of Learning*, Chapter XII, Part II, 41st Yearbook, National Society for the Study of Education, pp. 415-443.
10. Burlingham, H. H., and E. M. Juergenson. *Selected Lessons for Teaching Off-farm Agricultural Occupations*. Danville, Illinois: The Interstate Printers & Publishers, Inc., 1967.

11. Clover, Everett. "The Problem Method." *The Agricultural Education Magazine*, Vol. 26, September 1953.
12. Collings, Ellsworth. *An Experiment with a Project Curriculum*. New York: The Macmillan Company, 1923.
13. Cross, Aleene A., ed. *Vocational Instruction*. Arlington, Virginia: The American Vocational Association, Inc., 1979.
14. Crunkilton, John R. *Teaching the Disadvantaged: A Curriculum Guide for Classes of Disadvantaged Students in Agricultural Education Programs*, rev. ed. Richmond, Virginia: Agricultural Education Service, State Department of Education, 1975.
15. Crunkilton, John R., and Paul E. Hemp. "The Curriculum: Professional Education." *Teacher Education in Agriculture*, 2nd ed., Arthur L. Berkey, ed. Danville, Illinois: The Interstate Printers & Publishers, Inc., 1982.
16. *Develop a Course of Study*, Module 8. Columbus, Ohio: The Center for Vocational Education, The Ohio State University, 1978.
17. Durkin, Helen Elsie. "Trial and Error, Gradual Analysis and Sudden Reorganization, An Experimental Study of Problem Solving." *Archives of Psychology*, No. 210, 1936.
18. Gates, Arthur I., et al. *Educational Psychology*. New York: The Macmillan Company, 1950.
19. Glasser, William. *Schools Without Failure*. New York: Harper and Row, 1969.
20. Hammonds, Carsie. *Teaching Agriculture*. New York: McGraw-Hill Book Company, 1950.
21. Hildreth, Gertrude. "Puzzle Solving With and Without Understanding." *Journal of Educational Psychology*, Vol. 33, pp. 595-604, 1942.
22. Hoover, Kenneth H. *The Professional Teacher's Handbook: A Guide for Improving Instruction in Today's Middle and Secondary Schools*, 2nd ed. Boston: Allyn & Bacon, Inc., 1975.
23. *Horticulture: A Curriculum Guide for Agricultural Education*. Richmond, Virginia: Agricultural Education Service, State Department of Education, 1980.
24. Joyce, Bruce, and Marsha Weil. *Models of Teaching*. Englewood Cliffs, New Jersey: Prentice-Hall, Inc., 1972.
25. Krathwohl, David R., Benjamin S. Bloom, and Bertram

# BIBLIOGRAPHY

Masia. *Taxonomy of Educational Objectives, Handbook II: Affective Domain*. New York: David McKay Co., Inc., 1964.

26. Krebs, Alfred H. "Discipline—Problem and Opportunity." *The Agricultural Education Magazine*, Vol. 28, October 1955.
27. Krebs, Alfred H., ed. *The Individual and His Education*. Washington, D.C.: The Second Yearbook of the American Vocational Association, 1972.
28. Krulik, Stephen, and Jessa A. Rudnick. *Problem Solving–A Handbook for Teachers*. Boston: Allyn & Bacon, Inc., 1980.
29. Lancelot, W. H. *Permanent Learning*. New York: John Wiley & Sons, Inc., 1944.
30. Liggett, Kenneth J. "Problem Solving Procedures in Teaching Agriculture." *The Agricultural Education Magazine*, Vol. 23, pp. 185-186, February 1951.
31. Mager, Robert F., and Kenneth M. Beach, Jr. *Developing Vocational Instruction*. Belmont, California: Fearon Publishers, Inc., 1967.
32. Maslow, A. H. *Motivation and Personality*. New York: Harper and Row, 1970.
33. Morgan, Barton, Glenn E. Holmes, and Clarence E. Bundy. *Methods in Adult Education*, 3rd ed. Danville, Illinois: The Interstate Printers & Publishers, Inc., 1976.
34. *Natural Resources Management: A Curriculum Guide for Agricultural Education*. Richmond, Virginia: Agricultural Education Service, State Department of Education, 1972.
35. Pautler, Albert J., Jr. *Shop and Laboratory Instructor's Handbook: A Guide to Improving Instruction*. Boston: Allyn & Bacon, Inc., 1978.
36. Phipps, L. J. *Handbook on Agricultural Education in Public Schools*, 5th ed. Danville, Illinois: The Interstate Printers & Publishers, Inc., 1980.
37. Purcel, David J., and William C. Knaak. *Individualizing Vocational and Technical Instruction*. Columbus, Ohio: Charles E. Merrill Publishing Company, 1975.
38. Orlich, Donald C., et al. *Teaching Strategies: A Guide to Better Instruction*. Lexington, Massachusetts: D. C. Heath & Company, 1980.
39. Redfield, Doris L., and Elaine Waldman Rousseau. "A Meta-analysis of Experimental Research on Teacher Questioning

Behavior." *Review of Educational Research*, Vol. 51, pp. 237-245, Summer 1981.
40. Stewart, W. F. *Methods of Good Teaching*. Columbus, Ohio: The Ohio State University, 1950.
41. Sutherland, S. S. "A Problem Solving Procedure." *The Agricultural Education Magazine*, Vol. 21, August 1948.
42. Weaver, H. E., and E. H. Madden. "Direction in Problem Solving." *Journal of Psychology*, Vol. 27, pp. 331-345, 1949.

# INDEX

# ► INDEX ◄

## A

Adult plan
    beef production, 217
    crop production, 213
    farm management, 218
    indoor plants, 215
    outline for, 207
Adult teaching, 205
    classroom procedures, 213
    evaluation, 212
    group objectives, 210
    interest approach, 209
    problems and concerns, 211
    situation, 208
    teaching / learning objectives, 209
    teaching outline, 207
Agricultural business course, 289
    planning for, 131
Agricultural principles, 309
Aiken, 63, 65
Application
    adult, 212
    high school, 21
Approved practices, 22

## B

Brownell, 65

## C

Chemistry plans, 300
Classroom procedures
    adult, 213
    high school, 137
    seating arrangements, 244
Collings, 60
Completion questions, 26
Course building, 153
    list of agricultural topics, 265
    principles, 258

    seasonal considerations, 13, 201
    steps, 257
Course title, adults, 207

## D

Daily planning, 28
    daily planning sheet, 30
Demonstration outline, 317
Discipline
    conducting a conference, 182
    controlling problems, 180
    definition, 177
    preventing problems, 178, 246
    using physical force, 184
Discussion leading, 139-144, 148, 155
Durkin, 55

## E

Enterprise
    definition, 6
    examples, 8-12, 265
Essay questions, 27
Evaluation, 21
    adult, 212
    laboratory, 246
    teacher, 245
    teaching, 241
    teaching / learning activities, 231

## F

Fuller, 300

## G

Gates, 37
Group objectives, 18, 139, 210

## H

Hildreth, 57

## I

Instructional aids, 89, 90
    function of, 91
    selection of, 96
Interest approach, 16, 138
    adults, 209
    (*See also* Motivation)

## L

Laboratory instruction, 189, 246
    evaluation of, 246
Laboratory practices, 189
    clean up, 198
    evaluation, 201
Learning, concepts of, 35
Lesson plan format, 117

## M

Maslow, 73
Matching questions, 25
Minimum plan
    illustration, 161
    outline, 159
Motivation
    definition, 71
    extrinsic, 74
    intrinsic, 72
    Maslow's theory, 73
    motivational tips, 75, 239
Multiple choice questions, 26

## N

New developments, planning for, 130
Notebook use, 151

## O

Objectives
    action verbs, 97
    adult, 210
    behavioral, 96
    group, 18
    teaching / learning, 14
    trouble spot, 147

## P

Photography, plans for, 300
Placement-employment, planning for, 131

Plan, adult
    beef production, 217
    crop production, 213
    farm management, 218
    indoor plants, 215
Planning
    agricultural mechanics, 120
    daily, 28
    minimum, 159
    new development unit, 130
    teacher-student, 153
    for teaching, 3
    theory, 41
    young farmers, 226
Plan outline
    minimum plan, 159
    parts of, 4, 5
    by problem area, 4, 107, 109, 116, 122, 127
    understanding the plan, 5
    by unit, 112, 116, 125, 164
    using in classroom, 137, 213
Principles
    agricultural, 309
    course building, 258
    learning
        principle of association, 40
        principle of effect, 38
        principle of practice, 35
Problem area
    definition, 6, 7
    examples, 8-12, 265
    outline plan, 4, 107, 109, 116, 122, 127, 161
    relation to unit, 6
    trouble spot, 147
Problems and concerns
    adults, 211
    definition, 18, 139
    steps in solving, 19, 140
Problem solving
    other than agricultural topics, 297
    photography, 300
    steps, 19, 140, 151
    variations in, 169
Procedures, teaching
    adult, 213
    high school, 16

## Q

Questioning by the teacher
    analysis, 82
    comprehension, 82

# INDEX

evaluation, 82
evaluation of, 241
recall, 81
types of questions, 80
use of, 80

## R

Research on learning, 55

## S

Situation
    definition, 13, 208
    trouble spot, 147
Steps in solving problem, 19, 140, 151
Student participation, 233, 241
Student related to learning, 55
Supervised agricultural experience programs
    placement programs, 289
    plans for, 22
Supervised study, 20, 141

## T

Teacher
    evaluation of, 245
    importance, 251
Teaching / learning objectives
    adult, 210
    defined, 14
    high school, 18
    trouble spot, 147
Teaching procedures, 16, 213
Teaching resources, 28, 212
Teaching strategies, 88, 89
Teaching techniques
    definition, 89
    demonstration, 317
    evaluation of, 102, 231
    function of, 91

related to problem solving, 90
solution of, 94, 96
supervised study, 20, 141
trouble spots, 152
use of, 28
Teaching young farmers, 226
Tests, 23
    completion questions, 26
    criteria for construction, 23
    essay, 27
    matching, 25
    multiple choice, 26
    problem-solving questions, 27
    true and false, 24
Theory of planning, 41
Trouble spots in teaching, 147
True and false questions, 24

## U

Unit
    definition, 6
    examples of, 8, 9, 10, 11, 12, 265
    outline plan, 5, 112, 125, 164
Using lesson plans
    classroom, 107, 109, 112, 116, 127, 137
    laboratory, 120, 122, 125, 144
    placement-employment settings, 131

## V

Variations in problem solving, 169

## W

Weaver and Madden, 58

## Y

Young farmer teaching, 226